THE POWER OF Permaculture Principles

Wilf Richards

Published by
Permanent Publications
Hyden House Ltd
13 Clovelly Road
Portsmouth
PO4 8DL
United Kingdom
Tel: 01730 776 582
Email: enquiries@permaculture.co.uk
Web: www.permanentpublications.co.uk

Distributed in North America by Chelsea Green Publishing Company, PO Box 4529, White River Junction, VT 05001, USA
www.chelseagreen.com

Distributed in Australia by Peribo Pty Limited, 58 Beaumont Road, Mt Kuring-Gai, NSW 2080 Australia
https://peribo.com.au

Cover artwork and book design by Two Plus George Limited, info@twoplusgeorge.co.uk

Printed in the UK by Bell and Bain, Thornliebank, Glasgow

This product is made of material from well-managed FSC®-certified forests and from recycled materials and other controlled sources.

The Forest Stewardship Council ® (FSC) is a non-profit international organisation established to promote the responsible management of the world's forests. Products carrying the FSC label are independently certified to assure consumers that they come from forests that are managed to meet the social, economic and ecological needs of present and future generations.

British Library Cataloguing-in-Publication Data
A catalogue record for this book is available from the British Library

ISBN 978 1 85623 195 4

Praise for the book

Wilf, together with a wonderful community of collaborators, has created a book that sits beautifully at the intersection of designing and educating, and the heart of permaculture itself. *The Power of Permaculture Principles* offers a grounded, thoughtful, and inspiring lens through which to explore the principles – not just as ideas, but as guides to engage with and regenerate living systems. It supports designers to deepen their practice and helps educators bring more clarity, creativity, and care into their teaching. A truly valuable addition to any permaculture library.

Morag Gamble,
Permaculture Education Institute

Permaculture is a constantly evolving body of ideas, and navigating the sheer plethora of permaculture principles intended as aids to our design thinking can end up being confusing rather than helpful. Wilf and his collaborators have managed to pull together and bring coherence to a sometimes bewildering array of guidelines and precepts derived from many writers, thinkers, and practitioners – no small feat! What emerges is a distilled 'pattern language' of permaculture thinking tools, an invaluable resource for teachers, designers, or indeed anybody wanting to apply permaculture to their daily lives.

Graham Burnett,
Permaculture teacher and author,
including *The Vegan Book of Permaculture*

An extraordinary contribution to permaculture design thinking, this is a brilliant example of how people can learn from nature and become better earth citizens. A must read for anyone interested in becoming a more effective and ethical decision-maker. Wilf and his collaborators have made a generous contribution to expanding the potentials and possibilities of the global movement.

Delvin Solkinson,
Diploma tutor and coordinator, Permaculture Institute
Dean and Instructor, Permaculture School at Pacific Rim College

A well-written book providing a deeper theoretical understanding of the principles. We look forward to seeing it studied in Danish permaculture courses.

Mira Illeris and **Esben Schultz**,
Permaculture farmers and authors of
Permaculture: Practical and Radical Solutions for a World in Crisis

I greeted this book with relief because after nearly 50 years of permaculture we needed an in-depth scrutiny and throught-provoking book that explores the many principles that have arisen and transformed since Holmgren and Mollison. This collaborative book with a mix of authors and lead by Wilf Richards is timely and valuable. Although Euro-centric, I look forward to it stimulating many conversations globally.

Rosemary Morrow,
Permaculture teacher and practitioner

Principles and ethics are foundationally important for guiding all aspects of life, design and interaction, with ourselves, other humans, animals, all of nature and all around us, now and with the future in mind. Wilf has dived deeply into an exploration of permaculture principles and expands our thinking and provides a resource for us to learn and grow our practice with. I celebrate his gift to the permaculture community and encourage us all to actively integrate the many gems of wisdom into our life, lifestyle and livelihoods.

Robin Clayfield,
Permaculture pioneer, creative facilitator,
educator and bestselling author

Many a permaculture tutor has had the challenge of finding relatable examples of the permaculture principles to enlighten their learners, whether for land-based or invisible designs. By bringing thoughts from permaculture thinkers and doers together with his own experience, Wilf offers a comprehensive compendium of integrated voices to teach us all. As a prospective textbook it offers us new knowledge, with impressive references and notes. For me, it read like a well-loved novel, slipping in thought-inspiring images and questions, turning arcane language into 'permaculture slang'. And it's got a great plot…

Jan Mulreany MSc,
Diploma Permaculture Designer
and co-founder of Brighton Permaculture Trust

The Power of Permaculture Principles is a comprehensive and in-depth examination of the sayings, ideas, and guidelines that are part of permaculture's design system. Based and expanding upon Holmgren's and Mollison's thinking it offers thoughtful insights with practical considerations and applications that will benefit both beginners and long-time practitioners. It is a valuable addition to the evolving discipline of permaculture.

Rico Zook,
Permaculture designer, consultant and educator

About the Author

Wilf Richards is a permaculture designer, who has been co-managing a cooperative smallholding business near Durham City with friends since 2001. The ongoing learning about land, integration of community and creation of systems feeds into his permaculture teaching. He is a co-founder of the worker's coop Abundant Earth, The Land of Roots CIC and Green Durham CIC. As a Senior Tutor for the British permaculture diploma system he supports apprentices in developing their professional permaculture design skills. On the land he focuses on managing the market garden, the veg box scheme and volunteers.

I want to dedicate this book to Graham Bell and Marina O'Connell, two of my collaborators who died whilst this book was being written. They were both mind blowingly inspiring people who have contributed so much to the permaculture movement.

More from Wilf at: https://linktr.ee/wilf.richards

Contents

Introduction

I have always been curious that we don't have more books that focus on the principles and ethics of permaculture. In fact there has, until recently, only ever been one and that was David Holmgren's book *Permaculture: Principles and Pathways Beyond Sustainability* (2002).[1] It was a game changer of course. Many of the other permaculture books out there typically feature a chapter or section about the principles, showing us the many variations and evolution of them over the years.

Since Holmgren's book, there appears to be many teachers only using his set of principles, resulting in the loss of Mollison's originals, many of which still pack a punch and are still required for most permaculture curriculums. For example, I don't think we should forget about Mollison's principles related to ***functions and elements*** even though they are included in Holmgren's ***integration*** principle. I have always taught both in my Permaculture Design Courses (PDCs) as my learning about permaculture started in the mid 1990s and thus predated Holmgren's book. Websites about the principles tend to focus on one set or the other, and tend to also only explore each principle with a sentence or two, lacking the depth they deserve. The principles are guides for us on our permaculture journeys and they have far greater potential and nuance when considered deeply. They can dramatically shift our attitudes and perceptions. They could even change the world! You will find both Holmgren and Mollison principles here, and principles from many other teachers too.

As a Diploma Tutor for the Permaculture Association in Britain, I regularly receive designs for assessment and feedback. These designs show the wide range of applications, interpretations and developments of the principles. I have seen over the years a need for some further detailed writing about the principles and I started that process in my new post-Covid PDC by creating simple handouts for each of the principles. These inevitably became first draft chapters for this book.

> *The world is a tricky, complex place. There are not very many simple answers. It takes dedication, skill, and above all experience, to be able to manage it well. You can try to boil it all down into a set of simple principles, but these are all but useless to beginners, who invariably misapply them. The best set is undoubtedly David Holmgren's 12, but these are so profound you need a lifetime of experience to understand what they really mean.*
>
> **Peter Harper**[2]

I realised early on that there was no way that I could write a detailed book about the principles on my own, and I would not want to either. There is of course strength in hearing the ***diversity*** and actively increasing the ***yield*** through working together in ***cooperation.*** My immediate natural allies have been fellow Permaculture Teachers, Diploma Tutors, Diploma Graduates and Diploma Apprentices. I have worked with over forty amazing people to help bring this book together. I could not have done it without them and I am truly grateful for all of their time and enthusiasm for the project. They have brought their insights, knowledge and experiences to the book, and I have learnt so much as a consequence too.

I had been warned that taking an approach of including so many people could simply result in a dog's breakfast of a book. I hope you will disagree and will be as happy as any dog expecting its breakfast. I hope you will instead see the ***integration*** of many voices sharing what we think is key to developing a truly regenerative society that reunites us once again into nature's arms. This is not an anthology of a book but instead a range of collaborations facilitated and coordinated by myself as editor and the main writer. That range has included contributions, suggestions, edits, additions, interviews and conversations.

The principles are not carved in tablets of stone. Ask any two permaculturists what the principles are and you'll get two slightly different lists.

Patrick Whitefield[3]

In each chapter I have tried to include details about the origin of each principle and its journey through the years. Where relevant I have included definitions of keywords and how each principle is linked to other principles. As you may have noticed already, where a principle is named which relates to another chapter, it will be highlighted in bold and italics. I have also highlighted the permaculture ethics where they have been mentioned. In each chapter there may also be details of where the principle can be seen in natural systems and of course various examples of applications, both land based and non land based. At times there has been a need to provide some critical thinking about a principle, to check if it still works, the nuances of its use and various understandings or contradictions. There can, of course, be multiple opinions about each principle, and I have deliberately tried to integrate them all with my collaborators. It has always been important to me to not evangelise permaculture and to avoid believing in it too much. I think we always need to question our own belief systems. Let us remain curious about possibilities. As Robert Anton Wilson once said: "Belief is the death of intelligence."[4]

In many chapters there are also design tools that have been included, especially where they support the application and deeper exploration of a principle. At times, there are proposed tweaks to the principles and recommendations to balance them with other principles or ethics. I hope that I and the other collaborators have added clarity to the collective understanding of the principles. I hope this book will support your permaculture journey and will be useful to those new to permaculture and anyone fully immersed in its depths already.

Defining the Principles

Written in collaboration with Peter Cow and Tom Henfrey

What are principles?

The word *principle* can be defined as a fundamental assumption or guiding belief, as a scientific or moral rule, as a guideline or philosophical idea. It originates from the Latin *principium* which means beginning or foundation.[1] There are other words that have a similar meaning such as maxim, value, axiom, concept, idea, essence, truth or theory.

Principles are used by just about every discipline and belief system that exists, including religions,[2] law, governments, science,[3] engineering, medicine and the arts. The principles are normally a key fundamental part of a belief system. Religious principles are normally of the moral kind to guide our behaviour, for example they include kindness, charity, justice, truth, compassion, love and forgiveness. Religions also have cosmological principles to provide some sense of how reality came into being or how it works.

Science is awash with hundreds of principles such as Bernoulli's principle about fluid dynamics and the principle of relativity. Most of the scientific principles are qualitative and tell us why and how something happens, as compared to scientific laws and rules that are quantitative and usually come with some extensive equations to tell us what will happen. In medicine and law, principles are typically guidelines or ethics, directing the correct approach to take or influencing procedures.

Permaculture is a discipline and belief system with various sets of principles. One of the attractions of a belief system is that it offers a set of principles and guidelines for how to live and maybe even a community with a shared cosmology and a collective understanding of reality. Maybe you already believe in or use the permaculture principles or maybe you are totally new to permaculture. Maybe, like me, you have found them to be very supportive of the decisions you have made in your life and maybe you have some questions about them. There is a danger to over-believing whatever set of principles guide your life. What set of principles have provided a foundation for your life? Are you just blindly following them? Do they influence your behaviour for the better? Can they be developed and improved? Who wrote them, where did they come from and why do they exist?

The emergence of permaculture principles

It seems likely there was no intention to have a set of principles for permaculture from day one. They gradually emerged and developed, and continue to do so. Bill Mollison, co-founder of permaculture, shared his first proper exploration of the principles in his book *Permaculture, A Designer's Manual* (1988)[4] but he was clearly developing them from the early 1980s.

Permaculture practitioners see the principles as the underlying mechanisms as to why nature is so successful and as a guide to tweak our behaviour to be more aligned with nature. Permaculture loves asking why modern societies are so detached from the natural world and finding ways for us to reconnect with nature. Through the exploration of this question the principles have emerged. David Holmgren stated in his book that the principles 'can be derived from the study of both the natural world and pre-industrial sustainable societies and that these will be universally applicable'.[5]

> *The idea of a simple set of guiding principles which have wide, even universal application is attractive. Permaculture principles are brief statements or slogans which can be remembered as a checklist when considering the inevitably complex options for design and evolution of sustainable systems. These principles are seen as universal although the methods which express them will vary greatly from one place and situation.*
>
> **David Holmgren**[6]

The need to communicate and teach permaculture has possibly been another key reason for the emergence of the principles. The principles enable us to communicate the complexity of permaculture across many different climates, landscapes and communities. Many great examples of sustainability have been developed but the solutions are often site specific. The best solutions, for any given situation, will be influenced by the conditions found at that moment in time within the given space. That could include soil type, climate, community capacity and local knowledge. So the pattern is: *here is a good technique that worked in that location but don't copy it exactly, instead use these guiding principles that we used and use those principles yourself to work out your own solutions in your context.* The principles can be seen to act directly on our creativity, our problem solving minds – inspiring and empowering us to find our own solutions to our unique design challenges and not just copy what has happened elsewhere. One of Bill's numerous catch phrases was, I have been told, *vuja de* – the opposite of déjà vu. The phrase summarises the uniqueness of every moment. In every design situation we should be aware that we have never seen this before, in this place and in this context. So pay attention, observe and find a tailored solution, using the principles as a guide.

Some see the principles as a form of common sense or as innate learning and passed-down wisdom. We can also see them as essential for our survival as a

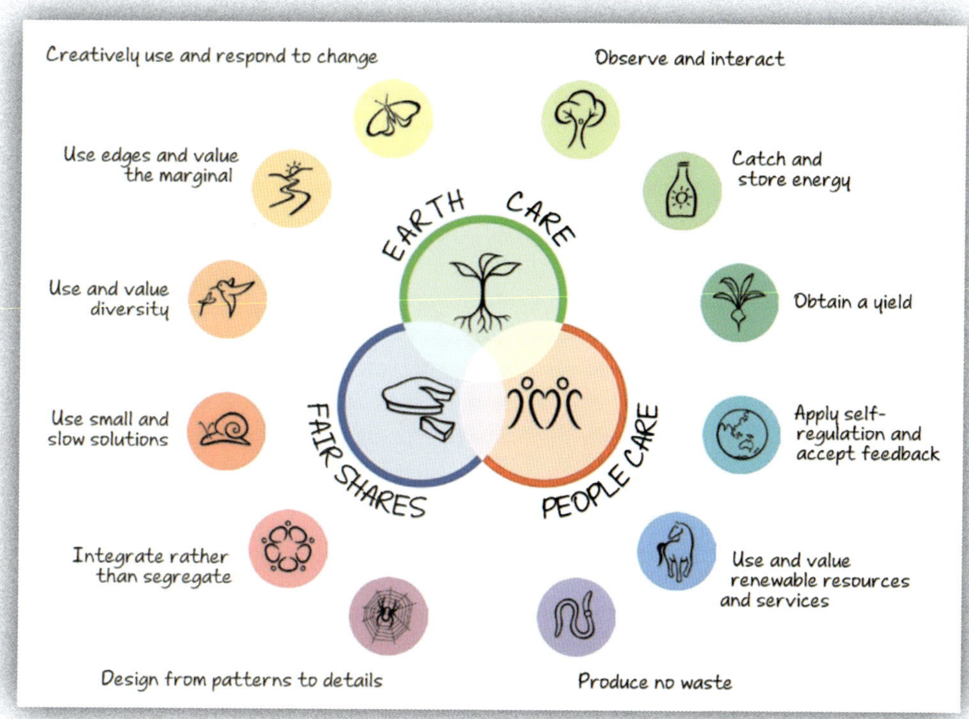

Holmgren's principles map © David Holmgren

species and the planet. The principles challenge our thinking and thus our behaviour too. Nature is complex, human society is complex, and we have a tendency to make decisions based on precedent.[7] The pattern is: *I did it like this last time, and that sort of worked ok so therefore I am going to do the same thing again.* Instead we need to question and evaluate our actions and shift our behaviour towards being more nature based and more ethical.

One definition of permaculture, used by many teachers, is *think like an ecosystem* and the principles are key to that shift. They encourage us to see beyond what we already know, to observe and consider both past and future resources, patterns and possibilities. They can support us to learn about and design for the physical and energetic relationships among the parts of any system. Rather than the linear thinking of the work and pollution modern mindset, they encourage us to see the ecosystem of connections that can be woven even more tightly and used to reduce the work we need to do and the pollution our actions and inactions cause.

Despite the mainstreaming of climate change action and sustainability, there is still a strong need for permaculture principles. They are holistic and provide a whole systems thinking approach. Commercial green action is often only focused on single issues such as dealing with carbon dioxide or increasing recycling.[8] In an attempt to achieve a green industrial capitalism it is easy to completely forget about what nature needs.

Indigenous influences

Permaculture has a huge debt to Indigenous Knowledge.[9] In the introduction to Chapter 2 of his book, Bill Mollison states: 'It is alarming that in western society no popular body of directives has arisen to replace the injunctions of tribal taboo and myth. When we left tribal life we left with it all guides to sensible behaviour in the natural world, of which we are a part and in which we live and die.'[10]

In many Indigenous and traditional societies, historical and present-day, behaviour may be regulated through custom, social norms, myth, ritual and taboo rather than by the rule of law or a centralised authority. The intention may be to highlight the consequences of actions to educate and lead people. Anthropologist Eugene Anderson refers to these customary mechanisms for social-ecological regulation as *Ecologies of the Heart*,[11] followed not due to coercion by any central political authority, but because they are embedded in people's basic ethical and cosmological understanding, in emotionally compelling ways. Permaculture principles are designed to guide behaviour in a similar way, and are not hard and fast laws. Many of the principles encourage us to behave, think and design in careful, wise ways – ways that match up with patterns of behaviour seen in indigenous cultures.

During his decade-long period at university, Bill Mollison also independently researched and published a three-volume treatise on the history and genealogies of the descendants of the Tasmanian Aborigines.[12] Writing of his encounter with Aboriginal Australian culture, specifically Pitjantjatjara women who inspired his understanding of the importance of ***edge***, he writes, 'In the Pitjantjatjara country, the celebration, the way, and the pattern are indistinguishable things – a totality.'[13] In his book *Sand Talk: How Indigenous Thinking Can Save the World*, contemporary Indigenous scholar Tyson Yunkaporta, of the Apalech clan in Queensland, Australia, describes five different ways of thinking that characterise indigenous knowledge: kinship-mind, story-mind, dreaming-mind, ancestor-mind and, most familiar to permaculturists, pattern-mind.[14]

Permaculture practitioners are, in some important respects, attempting to create new forms of indigeneity and often seek Indigenous knowledge as an ongoing influence. This can have a backlash when we don't get it right and can be perceived as cultural appropriation, even if there were good intentions behind the decision to use a great approach from another culture. David Holmgren, in his book writes, 'indigenous and traditional cultures of place have provided much of the inspiration, elements and design solutions, both in the original conception and in the ongoing evolution of Permaculture'.[15]

Ecologist Fikret Berkes has studied and collaborated extensively with Indigenous land managers across the world over several decades. He describes Indigenous knowledge as a *Sacred Ecology*[16] with multiple levels. In his model he places local knowledge at the centre, nested within management practices, then within social institutions (such as those for allocating usage rights and limits), and finally within a cosmology or worldview. It is within this last external circle of worldview that we would find principles.

The nature of the permaculture principles

We will see in detail in the following chapters how each of the permaculture principles evolved and emerged. For now let's just focus on the broad nature of all of those principles. Bill Mollison clearly states that the permaculture principles are not dogma. 'Dogmas are rules which are intended to force centralised control (often by guilt).'[17] He wanted us to be aware of the fundamental principles that govern natural systems and to use them as directives and guidance.

> *A principle is a basic truth, a rule of conduct, a way to proceed. A law is a statement of fact backed up by a set of hypotheses which have proved to be correct or tenable…Now I have evolved a set of directives which say: 'Here is a good way to proceed.' It doesn't have anything to do with laws or rules, just principles.*
>
> **Bill Mollison**[18]

I have asked many people in the permaculture community how they would define the permaculture principles and there are lots of words and phrases that emerge. The image below is an attempt to share those keywords that repeatedly come up.

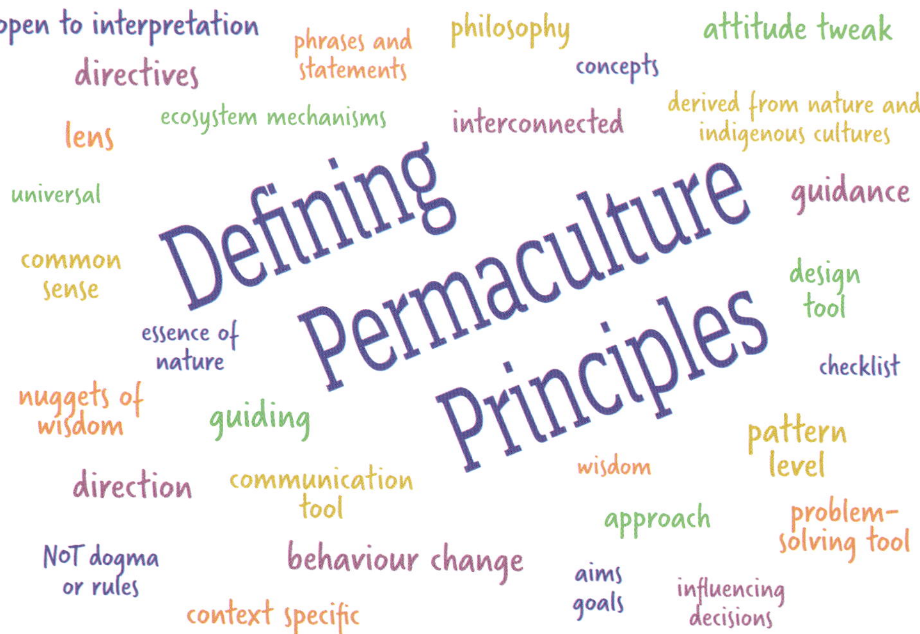

Keywords defining permaculture principles © Wilf Richards

The principles are not specific instructions such as 'do not do this' or 'always do that'. Permaculture principles are more open to interpretation, very transferable and context specific. They need to be applied carefully in consideration of the design goals and often in combination with other principles. They can also be over-applied and under-applied as we will see in later chapters. They can operate at both the ***detail*** level and the ***pattern*** level. How we use them and their application in specific contexts can be seen as the detail level but the broader saying and wording of the principle can be seen as a pattern. This also means that they can sound like aims, goals or aspirations at times.

There are many parallel movements similar to and overlapping with permaculture and many of them also have principles. For example, regenerative agriculture has a range of principles depending on where you look.[19] They tend to be at a more ***detailed*** level and some even sound more like rules, for example 'minimise soil disturbance' and 'reintroduce livestock' are very specific.[20] At the other end of the spectrum some of them include broader ***pattern*** level statements such as 'increasing biodiversity and ecosystem health'. Organic farming also has two levels of principles. First there are four principles that operate at the ***pattern*** level, as stated by the International Federation of Organic Agriculture Movements (IFOAM – Organics International). Those four principles are *health, ecology, fairness* and *care.*[21] Organic farming organisations then also have a second set of more ***detailed*** level principles which sound more like rules, such as limiting use of pesticides and herbicides.

One Planet Living principles, developed by the charity Bioregional in 2003, include big ***pattern*** level goals such as *health and happiness, equity and local economy, sustainable water* and *zero waste.*[22] They were used as the foundation to develop the United Nations Sustainable Development Goals, a universal call to action to end poverty, protect the planet and improve the lives and prospects of everyone, everywhere. We can see here how broader principle statements can easily also be seen as goals. Similarly biomimicry, developed by Janine Benyus,[23] includes both pattern level and more detailed principles.[24]

Some practitioners sort the principles into different types such as abstract principles and design principles.[25] This is often with the intention of categorising principles from those at the theory end through to those at the more practical scientific end. The permaculture principles, although often sounding quite abstract or theoretical, should always be applied within a practical and scientific understanding of our world. Permaculture practitioners, in their love of nature, generally have a deep respect for the scientific perspective. We want to know about gravity, the sun, water, soil, climate and all aspects of life. When the principles are applied then they can lead to, but are not the same as, detailed strategies and techniques such as solar power, diverse communities, forest gardens, swales, dams and the use of perennial crops.

What about the permaculture ethics?

Permaculture has two key foundational parts to its approach. They include the permaculture ethics as well as the principles. Most commonly the ethics are named as ***people care, earth care*** and ***fair shares*** but there are variations or subsections including limits to consumption, future care and sharing surplus. This book does not focus on the ethics but they cannot be separated from the principles. The ethics will be referred to in this book when they are clearly connected and important to a principle, for example principles around ***renewable resources*** are often signing us towards ***earth care*** and ***fair shares.*** In some cases I would argue that it is essential to include ethics when considering a principle, for example notions of increasing our ***yield*** need to be understood within the limitations of a finite world and thus considering ***earth care*** and ***fair shares*** too.

Without ethics, principles can be manipulated, so the two must be introduced together and ethics prioritised.

Rosemary Morrow[26]

Some permaculture practitioners will refer to the permaculture ethics as the ***ethical principles.*** They have similar characteristics to the principles in the sense that they can be seen as higher level goals. Many see the ethics as the core of permaculture, a deeper layer underneath the principles. Some see the principles as a way of moving towards the permaculture ethics. Ethics frequently deal with the complexity of morals and notions of right and wrong, similar to religious principles.

Using the Principles

Written in collaboration with Tomas Remiarz and Hannah Thorogood

Now that we have some sense of what a principle is, we can look at how to use them. There are of course many ways to use the principles at various stages in a design process.

Design processes

The focus of this book is not design processes but I need to explain some before we proceed. When you enter into the world of permaculture design you can get baffled by the abundance of jargon. All disciplines have their jargon as internal shorthand for ease of communication. Permaculture designers commonly use SADIM, OBRADIM and CEAP.

> SADIM stands for Survey, Analysis, Design, Implementation and Maintenance.
>
> OBRADIM stands for Observation, Boundaries, Resources, Analysis, Design, Implementation and Maintenance.
>
> CEAP stands for Collect information, Evaluate, Apply principles and Plan.

There are many variations of these design processes with new ones appearing all the time as we continue to develop our theory through practice. There is a pattern to all of the above design processes. They all go through stages of *observation* and survey work to see what is present already. Then they go through an *analysis* stage to judge what is present and explore options for change. Then there is the *design* stage where decisions and plans are made. Finally we get on with implementing and maintaining the system that has been developed. Once a design is established then we also need to *evaluate* it and learn from it, which may lead to tweaks or even complete restarts. This additional evaluation stage at the end can result in our jargon acronyms including an extra e at the end, for example, SADIME.

Survey	What principles can I observe in the system already? Which principles are present in this landscape? Which principles have already been applied?
Analysis	How might the principles support our judgements of what is going well and not so well on this site? How might the principles influence our options?
Design	Which decisions are best supported by the principles?
Implementation	How can I implement this plan using the principles? How are the principles manifesting as I implement this plan?
Maintenance	How might I improve this system by utilising the principles?
Evaluation	How did the principles help my design? Did I use any principles unintentionally? In what way could I have used the principles differently or better? What did I learn from using the principles?

Using the principles within the SADIME design process ©Wilf Richards

Using the principles in the typical design process

When we are making observations we may notice principles already in play. These could be simply nature in action or something implemented by the previous designer of the system that we have taken on. Maybe we notice the current ***diversity*** of plants on the site, the ***renewable resources*** that might be available to us or the ***resilience*** of the existing system. Maybe we notice a rainwater system implemented by the previous land user and we think of ***catch and store energy***, the compost system and we think of ***produce no waste*** or the lack of resources available and we think of ***limiting factors***.

When we are analysing our design challenge we may add judgements about what is working well and not so well with the current situation. This may bring up principles too, such as ***cooperation***, when we see how many happy people are already involved in that community garden that we are working at. Or maybe it is the organic gardening approach already adopted, showing how wildlife can be incorporated into our food crops, and we think of ***everything gardens*** or maybe ***working with nature***.

Typically in the second half of the analysis stage we start to explore potential solutions. We may see the opportunity to increase our ***yields*** by adding more ***edge*** growing to the site, or the possibility to improve an existing system through an ***energy efficiency*** idea we have. Given a choice of options we may choose one over the other because it demonstrates more principles. It may also tick more of the ethical boxes too. And so a key way in which permaculture designers move towards a design decision is by asking which ideas will fulfil the most principles and ethics.

When we are implementing our design decisions reality takes a hold, which of course is always slightly different from our theoretical plans and drawings. As the implementation unfolds we realise we can tweak our plans to fit the principles even more. For example, when I was rebuilding our compost bays[1] I realised that the timber boards that I was using could be cut in two different ways for different walls, and the length of one of the walls could be tweaked slightly resulting in less offcuts of timber, an example of ***produce no waste***.

Then finally, after finishing the implementation of our designs, we begin the long slog of maintaining our systems. And yes, hopefully our designs were so good in the first place that we have virtually no work to do now and we can just sit back and reap the harvest. But let's face it, usually we stumble upon an error, or something breaks down or we see a tweak that could improve the situation to save us time and money. Through evaluating our systems we can see another opportunity to put even more principles into action. For example, as part of a house energy design we could start by stripping out the gas boiler and replacing it with a wood burner with a back boiler to switch to ***renewable resources***. Then later on, after evaluating the design, we could add in solar hot water too, to add an additional ***element*** to our important ***function*** of heating our home, thus adding even more ***resilience*** to the system. Later still we could add more insulation, reducing the need for firewood, thus increasing ***energy efficiency***.

Can I use the principles retrospectively in a design process?

Yes you can, but you will learn so much more by considering them in advance and applying them as you go. The retrospective use of principles is common in those new to permaculture designing because there is a tendency to downplay or even skip the planning process and to get stuck straight into the excitement of action. If this sounds like you then I encourage you to reflect on where the principles were applied unconsciously and accidentally. As we shift towards being a more experienced designer we want to aim to use the ethics and principles consciously and deliberately, to help us shape our decisions as we go through the design process. In this way we can use them in layers with increasing complexity as they guide us towards our permaculture goals.

> *The more I contemplate the meaning behind each principle the larger each principle becomes. They twist and turn from one metaphor to the next seamlessly weaving layer upon layer to form a magical carpet that will take you on a journey through your thoughts.*
>
> **Stephen Andrews, Permaculture Diploma Graduate 2022**[2]

Using the principles in the Design Web process

In 2012, Looby Macnamara broke this pattern of the typical design process of observe, analyse and design through the launch of her Design Web process as featured in her book *People & Permaculture*.[3] This process changed the potential for how we frame our design challenges and how we use the principles in a design process. The process involves twelve anchor points arranged in a circle. You can start at any of the anchor points. If your project is new it probably makes sense to start at *Vision*, whereas an existing project means you are more likely to start at *Reflection* or *Appreciation*. One of these anchor points is clearly marked as *Principles*.

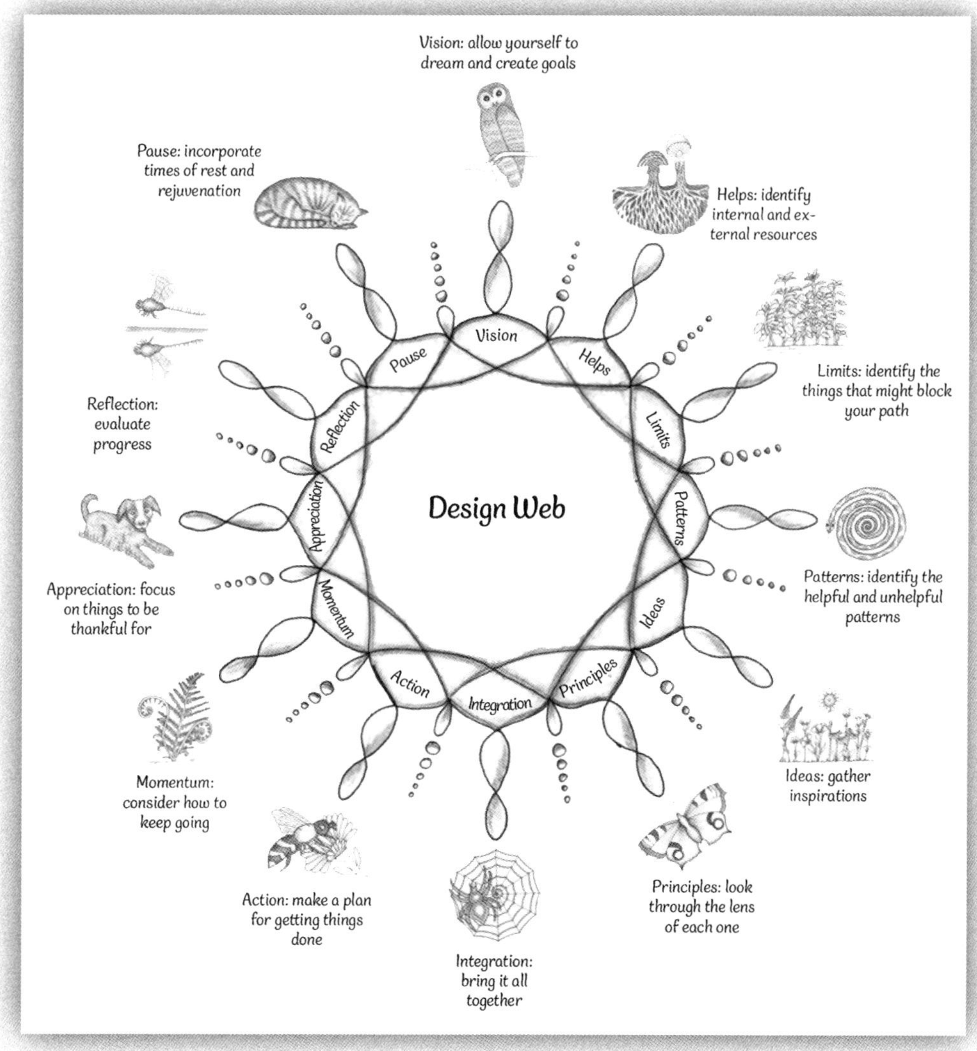

The Design Web includes principles, bottom right © Looby Macnamara

It is here that we are invited to focus our intention on the principles and look through the lens of each of them. We are of course not restricted to using the principles in only this location. They can easily pop up almost anywhere in the Design Web. I have seen ***catch and store*** in the *Patterns* stage, ***functions and elements*** in the *Integration* phase and of course ***patterns to details*** in *Patterns* and ***limiting factors*** in the *Limits* section.

Do we need to use only one set of principles at a time?

There have been various sets of principles developed over the years including those by Mollison, a group of twelve from Holmgren and developments into social principles such as those by Starhawk[4] and Looby Macnamara's Cultural Emergence[5] principles. I recommend that you explore what works best for you. Options could include sticking to one set of principles for a particular design challenge or intuitively choosing ones that speak to you. Many permaculture practitioners have tweaked the principles to make them more personally understandable and relevant to their situation and knowledge, such as Neil Kingsnorth's review in his diploma journey.[6] One particularly useful resource is Delvin and Grace Solkinson's Permaculture Design Deck, a beautiful set of cards that includes dozens and dozens of principles.[7] You could break out of your principle's comfort zone's by picking a few out randomly, from any or all of the sets in a similar way to a tarot card reading. I will use all of these approaches depending on the context. My personal favourite is to start with those that intuitively jump out at me as obvious for the situation and then to use a set, or more than one set, to explore what further principles could be in play that are less obvious that could shift my thinking.

Using the principles with other principles and with the ethics

As we will see in this book there are times when principles can contradict each other, support each other, be over-used, under-used or even not applied appropriately, especially if the ethics are not used alongside. This is a unique feature of permaculture, and one that keeps its integrity intact. In some cases this is automatic and obvious, for example, applying ***renewable resources*** is very likely to also be applying ***earth care***. We may see better results when we combine several principles together such as an increase in ***yield*** when we remove a ***limiting factor*** and increase ***energy efficiency***. Alternatively we may see ***integration*** over-used in a forest garden and some plants will have too little light and suffer. Principles are not separate entities, they are interlinked, influencing each other and providing multiple perspectives on a given situation. If we do not apply the ethics alongside the principles then we risk shifting away from our permaculture intentions.

For example, let's ensure we ***integrate*** with nature, applying ***earth care*** as we go rather than seeing the natural world as something separate.

The web of principles

In 2020, forest garden expert and permaculture designer Tomas Remiarz proposed a new way of using the principles, which he calls the Web of Principles. This involves using the Holmgren set of principles as the entire design process but using them in a similar way to Looby's Design Web. This is particularly useful for an intuitive quick design, making the use of the principles more conscious. Of course the process does not need to be limited to Holmgren's set of principles but could equally be applied to any set or even a mix of sets. The process enables connections between the principles to come to the surface more easily and provides a strong focus on the principles. Tomas found that he could also integrate principles from other sets, incorporate design tools and permaculture ethics into the process making it a deeper and more fully rounded design.[8]

One of Tomas's diploma apprentices, Neil Kingsnorth, tested the Web of Principles process in one of his diploma designs in 2022.[9] The topic was to identify the use of animals within his smallholding system. The evaluation section of the design includes reflection on the use of the web of principles process, which Neil found very conducive to and effective in decision making: *"The Remiarz Web of Principles uncomplicates, and clarifies, the design process. Instead of following a separate linear, or non-linear design framework and looking to apply principles as you go, the Remiarz model merges the principles and the framework into one."* The designs by Tomas Remiarz and Neil Kingsnorth using the Web of Principles can be seen in the design library on the Permaculture Association (Britain) website.[10]

An Introduction to the Principles

Now that we have explored a bit about how to define a principle and some ways in which they can be used, we can now dive deeply into each principle or family of principles. I say family because I am utilising many sources of permaculture principles for this book and many of the principles overlap or have been tweaked in wording over the years by various permaculture writers and teachers. I have tried to acknowledge everyone's input over the decades.

In each of the following chapters I try to cover a bit about the history, definitions, interpretations and the wide range of applications for each principle. I explore the links to other principles, connections to the ethics and in many cases include a relevant design tool. Each chapter is also written with my fellow collaborators who have added, tweaked and helped to edit the principle that they chose to focus on. No doubt more can be said about each one but we have tried to be succinct and focused. There are times when we question and scrutinise a principle and times when we add in cautionary tales of over-use and under-use. The principles are ordered alphabetically so as to not give any preference to one over another. The chapter titles are in some cases the name of a principle but in other cases hold the complexity of a family of many principles.

© Wilf Richards

Adaptation

Written in collaboration with Jen Rouse

This chapter is primarily influenced by the principle of ***creatively use and respond to change***, coined by David Holmgren. He shares details about the development of this principle in his book[1] and further discussion and examples appear on his website.[2] The only other place that this principle about adapting to change is found, from my research, is in Ian Lillington's book *The Holistic Life: Sustainability through Permaculture* (2007)[3] which tweaked the wording of the principle to ***use change constructively***. It seems that as permaculture designers we have not given much attention explicitly to adaptation and so here in this first chapter about the principles, I am going out on a limb to explore it.

Holmgren regarded his principle as an expansion of the earlier principle of ***accelerating succession***, which was coined and developed by Bill Mollison and Reny Mia Slay in their book *Introduction to Permaculture* (1991).[4] There is a separate chapter about succession later in this book as, to my mind, it deserves a large space of exploration all to itself, has already been quite extensively explored by numerous permaculture practitioners, and is quite different to adaptation. We will see how this principle, about our ability to adapt, has grown from historical spiritual perspectives and evolutionary theory, and is key to survival strategies.

I feel that the core intention behind this principle is to be aware that the world around us is in constant change and that we need to adapt to that changing world using our creative design skills. This could be to maintain an equilibrium or seize an opportunity. I often wonder why we do not have a principle clearly with the word *adaptation* in the title. After all, it has been said (and not by Charles Darwin) that "it is not the strongest of the species that survives, nor the most intelligent that survives. It is the one that is most adaptable to change."[5] So let's explore change and adaptation and see what that means for us as designers.

Impermaculture

Awareness that we live in a constantly changing world is key to understanding and working with this principle, and to being a healthy human. I love the Buddhist viewpoint of impermanence,[6] which clearly states that everything is constantly changing. Whether that be growth, decay, expansion, contraction, movement, birth, digestion, death or something else. These changes are on every scale and at every speed. We perceive many of them and yet most remain invisible to us, such as those at the chemical or microscopic levels or those imperceptibly slow to us.

Nothing in the known universe is fixed and these many moving parts lead to exciting emergent properties and enormous possibilities too. I have frequently thought that permaculture would have been better named *impermaculture*. We are not trying to create a permanent static culture that never changes. Instead we are interested in the play between the ***elements and functions*** over time, trying to constantly adapt to the ever changing world around us and ensure that we leave no dents on this special one-off planet.

Continuous change: that cloud is not there anymore © Wilf Richards

The relevant Buddhist teaching says that 'as long as there is attachment to things that are unstable, unreliable, changing and impermanent then there will be suffering'.[7] Notions of non-attachment are hard to grasp when we have a planet and the majority of species under threat from the crisis of climate and ecological collapse. We want to save it and keep it just the way it is, or rather was and will be, and yet we also need to adapt to the changing world around us. Adaptation is really important to us right now. We all need to use our creativity to come up with some very smart solutions to these problems and scale them up very quickly. No pressure. Remember the process can be fun too, indeed it needs to be fun because all tasks will never be finished and will never be perfect. We just stop every now and again and say that will do for today.

The key to survival

One of the core mechanisms of evolution found amongst all species is adaptation.[8] There are many types of adaptation according to biologists, occurring at various scales from long-term evolution to inherited changes within an individual.

Biologists do not regard learning, climatisation or flexibility as forms of adaptation as these are not inherited traits and all occur within the lifetime of a species. But these could all be precursors to long-term changes within a species or an individual and they are certainly part of our cultural understanding of adaptation. I will be using this broader understanding of the term adaptation for this book as we need to focus on things we can change. It is also useful for us to understand more about how adaptation occurs in natural systems.

Typically the first thing that happens to trigger a possible adaptation is a change in the environment of a species. This could be changes in the landscape, climate and weather such as ***ecological succession***, temperature or rainfall, pH or changes to another species in the environs. This in turn could trigger changes in behaviour such as needing to find different food or moving to a different habitat. These might be within the realms of the species' current ability to be flexible. Many species, including ourselves, are generalists and so we have a wide range of habitats we move through and settle in. Meanwhile other species, such as pandas, are specialists, largely only eating bamboo. When a species can change its behaviours, these could become new learnt habits or result in a permanent shift to a new habitat in which the species may adjust to suit that new climate. These changes may trigger mutations or changes in expression in the gene code of that species and the new behaviour might then become an inherited adaptation for future generations. Equally, mutations in the gene code could be triggered by a wide range of random environmental factors and not by changes in behaviour. All mutations may have a positive or negative effect on the species concerned. Of course only those mutations that benefit the species are going to be the ones that survive. This is where adaptation starts to occur at the evolutionary level. These long-term adaptations might result in new inherited behaviour, new physical traits or the repurposing of an existing trait, such as the insulating feathers of dinosaurs becoming the means of flight in birds. Species may adapt to each other's tweaks to continue ***integrating*** and maintaining a niche, such as the ongoing syncing of pollinating insects and flowering plants over millions of years. These changes all add to the ***diversity*** of planet earth.

An example of biological adaptation is that of the peppered moth[9] as noted by biologists since the 1800s. The moth comes in a variety of peppery shades from black through to white. In the early 1800s, the white variety was dominant across the whole of the UK. As the Industrial Revolution unfolded, the amount of sooty pollution on trees and buildings increased, especially in cities, and by the end of the century the darker variety of peppered moth was dominant. The blacker variants had an increased likelihood of survival due to better camouflage. They were less likely to be seen and eaten by the birds that predated them. The only places where the whiter variety was still dominant were in unpolluted rural areas. Then during the second half of the 20th century urban industrial areas started to clean up and the white variety started to get an advantage once again. This example of adaptation focuses on the change in the environment triggering the natural selection of a species. The moths had little choice in the matter.

Many animals do have opportunities to change their behaviour. For example, the reintroduction of wolves to Yellowstone Park in 1995 resulted in a shift of behaviour in the elk who started to avoid areas where they were more likely to be hunted.[10] This localised reduction in grazing pressure enabled trees such as aspen and willow to recover and the small bird population increased too. This in turn supported the growth in numbers of the beavers who used those trees to build their dams, which then led to changes in the river. All this from an adaptation in behaviour by the elk triggered by the reintroduction of the wolves.

The times they are a-changin'

As sea temperatures change, marine animals are shifting behaviours and location. In Scottish waters cod are moving north to colder waters and octopus, a warmer water species, are moving into those traditional cod locations.[11] The octopus have been developing new tricks in their new locations, learning to access the bait and lobsters in lobster pots. Meanwhile other species, such as beavers, adapt their environment rather than themselves, to better suit their needs. Humans have been doing this too for millennia, ever since we developed cities. And some species struggle to adapt at all, such as birds with relatively small brains, that are more likely to be killed by cars, whilst those with larger brains are showing plenty of signs of adapted behaviour to dodge the traffic.[12]

Given our current climate and ecological crisis, adaptation is key to the survival and ***resilience*** of all species. Those that can't adapt are likely to become extinct on a local level or possibly even globally. The speed of climate change is a major threat to all species on this planet, with changes in temperature too fast for some species to adapt to. Some species are keeping up with these changes and shifting to new locations or times, whilst those that do not change quickly enough are becoming very vulnerable. ***Small and slow*** solutions are not fast enough at times. The fastest to change are the microorganisms, especially bacteria. We know this all too well from overuse of antibiotics resulting in drug-resistant strains of bacteria, and the similar situation of overuse of pesticides and herbicides creating resistant bug and plant species.[13] As long as a few individuals within a species already carry a resistance to those chemicals then it will be those that survive and breed.

Tree leaves, flowers, seeds and insects have changed times of emergence, typically earlier in the season or for longer periods of time or even twice, once in spring and again in autumn. As gardeners and farmers this has pros and cons. From talking to community garden managers in southern UK there is evidence that walnuts are showing signs of increased likelihood of self propagation and shifting themselves north. At the same time fungal diseases are spreading and thriving in the warmer wetter conditions attacking a range of trees, including larch in the UK. This is one of the reasons we adapted our cooperative business, Abundant Earth, to include sawmilling of larch logs into timber for builders. The sex of alligators and turtle eggs are temperature sensitive. If the eggs are incubated above 34°C then mainly males are produced, while incubation lower than 30°C results in mostly females. Some species are starting to lay eggs earlier in the season and this is maintaining a balanced sex ratio.[14] The speed of climate change is pushing this ***edge*** and it may go beyond the point of where the alligators and turtles can keep up with environmental changes.

Some changes we would assume are beyond a species' capability but we may be wrong about that. Some ancient oak trees in Germany, over a thousand years old, have remembered their Spanish origin by changing their leaf shape to an ancient genetic variation that was still lingering in the tree's DNA.[15] This older leaf type is more resilient to extreme temperatures and droughts.

The power to adapt

Human evolution in the Great Rift Valley of Africa is a fascinating story of adaptation. This geographical area that runs almost the entire eastern length of Africa has a long history of dramatic environmental changes due to the underlying faults in the earth's crust, which cause mountains to form, volcanic activity, rips in the landscape and lakes expanding and contracting. These changes have been steady and constant for millions of years and are still happening now. This constantly changing environment speeds up evolutionary changes in the species that live there. Over the last three million years various *Homo* species have evolved in the Rift Valley and in many cases have moved out of Africa to spread north and east, including *Homo erectus* around two million years ago through to the latest being *Homo sapiens* between 60,000 and 200,000 years ago.[16] More than likely the reasons for humans to leave Africa each time are drought and starvation. We have adapted and survived by moving to greener pastures north.

Adaptation has been the key to *Homo sapiens*' success so far, and it will be a crucial requirement for the future of our species and the entire planet. In order to adapt, we first have to make ongoing ***observations*** of the changes in our environs. This is not just about climate change but any change in our circumstances. There may be a phase of denial before we accept this feedback of change. We then need to have a deep understanding of why it would be to our advantage to adapt to those changes. This in turn should lead to motivation to change and then we can design the best adaptation that we can make. Abundant Earth has adapted to various changes over the years such as changes in income, illness, ageing and going over the threshold of VAT. Our model of shared income has enabled flexible adaptation providing ***resilience*** too.

We also need to acknowledge that we can only change certain things. We are ***limited*** by our biology, capacity, awareness and resources. We are also strengthened by our imagination and creativity and our ability to move. We need to know what we can control and accept what we cannot change. As the Stoics taught, the only things that are in our control are our actions and our responses to our thoughts.

The jack of all trades

Being a generalist is powerful when it comes to adaptation. Evidence shows that specialists find it hard to adapt.[17] If your survival depends on only eating one source of food then you will be scuppered when that singular source of food is no longer available. Even as humans we are unlikely to only purchase from one shop and nowhere else. This is where the principle of adaptation overlaps with the principles of ***functions and elements***. If you have more than one skill or more than one income stream and more than one customer then you are more likely to adapt and switch roles as one of your work streams declines. 'A jack of all trades is a master of none, but oftentimes better than a master of one'.[18] There are pros and cons to being a jack of all trades. Specialise and you will develop mastery through that ***limiting factor*** but there is a balance to be had and ***diversity*** really helps with your ability to adapt.

Work with nature

Every day, in most locations around the world, the seasons are tweaking bit by bit, rotating through winter, then spring, then summer and autumn. There are of course many variations of the extremities of these seasonal changes: from the equator where these changes do not occur, through to the poles where the light changes of winter and summer are as extreme as they get. Some animals hibernate to cope with these changes. Many trees shed their leaves in autumn and are highly specialised in when they flower to suit both climate and pollinator. We can work with our local natural seasons and adapt our strategies through the year. We can see this most clearly when we grow food, learning when to sow crops and when to harvest.

In his forest garden in Devon, Martin Crawford noticed his local heritage apple varieties were no longer producing as well as they had in previous decades.[19] Warming winters mean that the old Devon varieties of apple weren't getting enough 'chill hours' to fruit. To respond to this, Martin has been experimenting with grafting apple tree scions from mid-France, where the climate is warmer, onto his existing rootstocks. ***Observing*** carefully and using the resources he has, Martin has ***creatively responded to change*** and this will hopefully result in more apples in the future. A similar change has been noted in the rhubarb triangle of Yorkshire where fewer cold days are reducing productivity with some growers turning to chemicals to maintain the output.[20]

Plants are sensitive to the increases in floods, heat waves and droughts. Thirty degrees centigrade is the typical stress point for most plants. Above that leaves may wilt due to dehydration or become bleached or scorched due to sunlight and heat. The chlorophyll in the plants will start to break down and transpiration of water from the leaves may exceed absorption of water by the roots. In extreme flood conditions plants may drown or at the least manifest yellowing wilting leaves, and there will be ***limitations*** on growth. Climate change is affecting our food crops dramatically and we need to have ongoing ***observations*** to note how our crops are being affected and adapt accordingly. Farmers and food growers have been masters at selecting species most suited to their specific climates in order to maximise yields for millennia.

There are many adaptation strategies that we can adopt to continue growing our food crops. We may need to increase the ***diversity*** of our crops and potentially look for new varieties that have certain characteristics, such as drought tolerance. In extreme heats we can limit evaporation of water from the soil with mulching and an increase in the quantity of water given to plants will help too of course, especially if you can get the water deep down into the roots. The extra heat in the UK summer may give us the opportunity to grow more plants outdoors, and increase the number of Mediterranean and frost sensitive plants, such as tomatoes, peppers, sunflowers and beans.

Changes to the landscape are an additional strategy that we could adopt. Swales will become increasingly important in areas of drought and sudden high rainfall so that the excess rainfall can be held back to minimise drainage, store water in the soil for longer and minimise flooding downstream. On a larger scale we have seen the use of rewilding strategies, such as the reintroduction of beavers, and engineering strategies, such as the construction of leaky dams, both having the ability to minimise flooding risks.[21] On a smaller scale, in your home garden, you might be increasing rainwater harvesting and developing raised beds to grow crops with minimal compaction and lots of organic matter to maintain a healthy more ***resilient*** soil.

As well as these longer term and seasonal adaptations we can consider the daily adaptations that we make. Every day we are influenced by changes in light levels from sunrise to sunset and through the night. We naturally adapt to getting up with the sun and sleeping when it's dark when our sleep hormone melatonin is functioning well. These biological daily rhythms are key to our wellbeing as they influence the amount and quality of sleep we get. Even when we shift our clocks from summer time to winter time or vice versa it only takes a few days of feeling confused and then we are back on track. Modern technology has had a big effect on these daily ***patterns***, from the invention of the light bulb through to the internet and mobile phones. Technological changes present us with another factor we need to consider in our adaptation strategies. In order to maintain our personal wellbeing we need to consider the pros and cons of each of these new tools as they emerge and use them wisely. We need to consider adaptation strategies for all areas of life when it comes to climate change: transport systems, housing design, land use, businesses, even politics. And nature has many solutions to share with us.

Catch and Store

Written in collaboration with Emma Leaf-Grimshaw and Pia Castleton

A key to resilience

The principle of ***catch and store energy*** is credited to David Holmgren as part of his set from his book.[1] It was previously mentioned briefly in Bill Mollison's *Designers' Manual*[2] in reference to damming water on contours. This is one of the principles that is key to ***resilience*** as the intention is to gather excess energy when there is an abundance, giving us a reserve to call upon when that energy supply is scarce and hopefully seeing us through to the next time of abundance. Permaculture cannot take credit for creating this principle as similar principles go back a long way, such as 'making hay while the sun shines' (from 16th-century English medieval farmers),[3] 'saving for a rainy day' (from an Italian play written in the 16th century)[4] and the more recent notion of a 'safety net' (probably from the mid-20th century). They have very similar meanings about taking advantage of the temporary abundance and to provide security in the future when that abundance will be lacking. For most of us that simple notion might just involve having a savings account to store your excess earnings, but where else can we use it and how is it commonly used by permaculture designers?

Catch and store in natural systems

We can see many examples of this principle in action in natural systems, such as plants and animals storing sugars; animals storing fats; animals stashing food items;[5] succulent plants storing water to cope with the desert conditions they live in; and soils storing water.

Plant photosynthesis at the most basic level of understanding is simply the capture of solar energy, carbon from the air and water from the soil to be combined into sugars that can be stored. The plant may use these sugars for daily requirements, such as respiration or ***cooperative*** trading with fungi. If there is excess then plants have a range of strategies depending on the species, such as storing sugar as starch in tubers, investing in the next generation as fruit, seeds and nuts, or developing the plant's infrastructure through the laying down of growth rings in the trunk of a tree. Perennial plants will catch and store carbon in their growth.

Animals store sugars in a different way. Either it is converted to glycogen to be stored in the liver and then released again to maintain blood sugar levels, or it can be stored as fat under the skin as a dense high-energy reserve, especially important for hibernating animals such as bears or those in food-scarce areas such as camels in deserts. Bees will gather nectar to convert and store as honey to feed their colonies over winter. Many animals will store their food harvests for a later day, such as squirrels burying nuts, seeds and berries; spiders paralysing prey to be eaten later; shrews using venom to immobilise their food to be eaten days later; jay birds storing acorns;[6] and moles hanging worms in their larders.

Types of energy we can catch and store

The word energy in this context can be interpreted as any material, whether physical or abstract, that can be retained: for example this could include water, food, energy, information and ideas.

Water can easily come in floods or not at all in droughts. Water can be stored easily and is absolutely essential for all life. Without it we have no ***resilience*** in any system. We can consider setting up rainwater harvesting systems for our gardens or building swales or dams to slow down the movement of water across the landscape.

Plants for food and medicinal uses can be very seasonal. We can wait for them for ages and then they can all arrive at once, such as fruit and nuts. Over thousands of years we have learnt to preserve gluts of food in a variety of ways including salting, drying, smoking, pickling, clamping,[7] cooking, canning, making chutneys, jams, vinegars and cheeses, and freezing. The oldest preserving technique is probably the ferment, where we have ***observed*** and developed processes involving a

Original rainwater harvesting set up in Abundant Earth's garden © Wilf Richards

symbiosis with various bacteria, yeasts and fungi to change the nature of our foods. The foods can be stored for a longer period of time and they typically have health benefits too, especially for our gut microbiome. All of these processes can keep us and our communities supplied until the following year or maybe even longer. If you are unfamiliar with these storage processes then I recommend getting a book on preservation and fermentation techniques, such as one by Sandor Katz.[8]

Plants may also provide us with medicinal parts that can be harvested and dried or preserved for future use. Our food cropping plants may supply us with seeds to save and store for future planting. Knowing what to harvest and when is a skill worth developing. We may use the leaves of the plant in the summer, seeds in the autumn or the roots in the winter. We may consider utilising plants that we weed out from our veg beds such as dandelion roots, which can be dried, roasted and used as a coffee substitute. Pruning a fruit tree could also be a gathering of sticks for a fire. And of course rather than seeing weeding as a chore, see it as a process to catch and store the fertile energy to use on your crops at a later date after composting. A lot of food is wasted and there may be opportunities to redirect these items into preserves or meals, such as the extraordinary work that waste-food cafés carry out. If we have the time then this principle can work alongside the principle of ***produce no waste*** at the same time.

Some plants act as nutrient or fertility accumulators. A great example, commonly planted on permaculture sites, is comfrey, that accumulates plant nutrients, especially potassium, and can be used to create a liquid fertiliser for use on flowering crops, such as tomatoes. Other examples include nettles that are

Comfrey plants and the liquid fertiliser created © Wilf Richards

high in nitrogen and normally grow in areas where there is excess nitrogen in the soil, and docks, with their very deep tap roots bringing up all sorts of nutrients and trace minerals, which can be cut repeatedly and the leaves added into the compost system. Plants can also be a nutrient and food source to herbivores and our livestock, for example the seasonal summer harvesting of hay is catching and storing that energy for later use as winter feed for the animals. This also lowers the fertility in the field which increases the potential ***diversity*** of wildflowers, an example where applying one principle naturally leads to another.

Plants can also provide fuel, typically in the form of firewood. Through their ability to catch and store carbon, we can have a ***renewable resource*** too. This could be done in an unsustainable way such as clear felling ancient mature woodlands: not great. Instead we can coppice selected trees, such as hazel and ash, whilst they are still quite young. This involves cutting the tree to the ground typically before ten years of growth and ensuring the tree can regrow new shoots, which given the right conditions it will with ease. This will provide us with a steady supply of firewood and material for other activities too, such as crafts. Firewood could also be sourced as waste wood, from tree surgeons for example.

Electrical power can be converted and stored in batteries which could be chemical in nature, such as lithium or lead acid batteries, or reservoirs of water, such as at the pumped hydro power station at Cruachan[9] in Scotland. These types of power stations have two water reservoirs. They use excess electricity in the power grid to pump water uphill to a top reservoir. Then when they need to supply electricity to the grid, the water in the top reservoir can be released to a lower reservoir via a turbine to turn the stored water back into electricity. On a smaller scale in an off-grid home, the sunlight can be captured via solar panels during the day, stored in batteries and then used in the evening.

Power from various sources, ideally ***renewable*** sources such as solar or wood, can be stored in the form of hot water, and if this is well insulated it can be used later. This is normal for houses with back boilers in their stoves, solar hot water systems or devices that store excess electricity as hot water. Heat generated in houses can also be stored in thermal masses, for example by the careful placement of masonry or bricks near to the source of the heat.

New neighbourhood-level heating systems have been developed in Finland by Polar Night Energy[10] whereby excess solar and wind energy is transformed into heat which is stored in sand. This is proving to be highly ***efficient*** and even more effective than storing the power in water, lasting for weeks rather than days, and reaching temperatures of 600°C.

Another elaborate way of storing heat has been developed by Jerome Osentowski and architect Michael Thompson at the Central Rocky Mountain Permaculture Institute.[11] Jerome farms at 7200 feet (over 2000 metres) above sea level in Colorado. There he has developed greenhouses with underground pipes and fans that enable excess heat generated in the greenhouses to be pumped underground where it is stored in the soil. It can then be released back into the greenhouses on colder days and at night to maintain fairly steady temperatures.

He refers to this as a 'climate battery' and it means he has been able to develop gardens inside the greenhouses far beyond the range of temperatures that the outside would allow, allowing the cultivation of tropical and mediterranean crops.

There are more abstract things that we can also catch and store, such as stories, music, information, images and ideas. The revolution of being able to write is thought to have emerged from the need to trade and record the number of items exchanged. Before this and in many cases still today, the original method of catching and storing these types of things will have been memory and learning, aided by recalling over and over again. These days we may use a notebook or journal to catch our stories and creative ideas for a project, or an audio recording of a conversation or a video of an event we attended. Maybe we need to store data for databases, information on websites, photos in an album, documenting a project or even writing a book, thus enabling the information to be retrieved and looked at again and again at a later date. Maybe we wish to catch and store positive feelings by using a memory jar[12] or a gratitude journal.

Powering up the resilience

When we add in the principle of ***all important functions supported by many elements*** then we can have a powerful, resilient system. For example, catching and storing rainwater in multiple locations and in multiple ways will dramatically increase our resilience to a drought situation. Why just have a single large tank when you could have multiple smaller tanks and also ponds, wells and dams too. If you add plenty of mulch and compost to your soil, and avoid compacting it, then the soil will store water too.

If someone accidentally leaves the tap of your single large tank on or it develops a leak, then no matter how successfully you have managed to catch and store an abundance, you may find it is all gone the following morning. We can strengthen our ability to catch and store by seeing what is holding us back from maximising our storage. Perhaps we have leaks in our systems that are acting as a ***limiting factor***. There is no point trying to gather water in a leaky barrel.

Not becoming greedy

So how about we just catch and store everything that we can? How about saving up all of the money that you earn or all of the rainwater that lands on your plot, or all of the food that we grow or buy. This is where this principle gets out of hand. We need to also consider the ***fair shares*** ethic, in particular ***sharing our surplus***. We need to allow for a flow of the energy and encourage sharing with those that need it. This could involve giving to charity, spending some of your spare time as a volunteer, sharing tools or the knowledge that you have accumulated. It might also lead to developing a business so that you can sell your excess. When you have an abundance then share, don't just catch and store.

Cooperation

Written in collaboration with Jenni Brooks, Barry Jones and Beth Silverbirch

Cooperation, not competition

It was Bill Mollison that first mentioned the principle of cooperation in his *Designers Manual*: 'The principle of cooperation: ***cooperation, not competition***, is the very basis of future survival and of existing life systems.'[1] David Holmgren in his book[2] extensively explored aspects of cooperation and competition, pointing out the complexities and dynamic tensions between these two forces. He brought aspects of cooperation and competition into his principle of ***integrate rather than segregate.***

> *Effective design for co-operation depends on a broad understanding of the functions of co-operation and competition. Simply deciding that competition and conflict are bad will not help us in this task. To the contrary, we need a balanced view of the contribution of all relationship dynamics to natural and human systems.*
>
> **David Holmgren**[3]

Other permaculture designers have also explored this principle around co-operation, including Colleen Stevenson using the phrase ***embrace cooperation,***[4] Sarah Spencer's ***cultivate cooperative relationships,***[5] Looby Macnamara's ***use the intelligence of cooperating hearts***[6] and even the simplest set of only five principles by Meg McGowan[7] included ***cooperate*** as one of them.

Cooperation and competition

At the British Permaculture Association's Diploma Gathering in March 2020, I witnessed competitive cooperation. It was fun but led to a far deeper thought that led me to further question the principle of ***cooperation, not competition.*** What I witnessed was the final presentations from Jenni Brooks and Barry Jones to complete their Diploma in Applied Permaculture Design.

About 18 months earlier Barry and Jenni had formed a 'response partnership'. Through this partnership they had regular check-ins, inspired each other and egged each other on. There was a friendly bit of competition between them and it was this striving together that led them to complete their diplomas at the same time. At times during the final leg one would find out that the other was a bit ahead and that would spur them on. It clearly started to become a friendly race to the finishing line, knowing that they must cross the line at the same time. They created mutual momentum, pushing and pulling each other along.

On the day, Barry presented first but I ensured that Jenni got her certificate first! It was beautiful to see a balance between cooperation and competition. At the end of Jenni's final presentation I proposed that we tweak the principle to ***cooperation and competition.*** It was a bit of a joke but something quite serious was brewing and it became one of the starting points for this book. The permaculture principles are an attempt to distil why natural processes are so successful. Both cooperation and competition exist as part of natural processes. So why have we, as permaculture designers, favoured one over the other?

In nature

There are plenty of examples of competition in nature, such as fights over territory, mating rights between animals of the same species, plants competing for light, and male birds singing enticing songs to attract the females attention and warning other males away. Over 50% of living organisms are parasitic in nature, the most extreme form of competition in nature, where one species benefits whilst there is an absolute loss by the other species. In some cases this can result in their death and in other cases the parasite will keep the host alive. Parasites can cause diseases such as malaria or live in the gut as do various parasitic worms. They also include a vast quantity of ticks, lice and many types of fungus.[8]

There are also lots of examples of cooperation in nature, such as mutual grooming to remove ticks, packs of wolves hunting together, and crows collaborating to push big birds of prey, such as buzzards, out of their territory. The research that has brought light on the relationships between trees and fungi in recent years has been outstanding, showing evidence of communication as well as resource sharing. We now know that trees can communicate to each other through these fungal mycelium networks using various message chemicals to warn of disease or predators.[9] The debate between whether cooperative or competitive forces are stronger in nature has been going on for years. Kropotkin, the Russian anarchist and aristocrat, wrote about mutual aid in the late 1800s.[10] He argued that animals are more cooperative than stated by the natural selection theorists, such as Charles Darwin and Thomas Huxley. Huxley described nature as constantly fighting, on par with a gladiator show, where the strongest survive and nature is nasty. The survival of the fittest theory does exist; we can see it when a cat takes out the slowest and weakest bird, but it is only one perspective. Kropotkin argued that the cooperation and mutual aid he saw between animals was evidence of empathy and morality in nature.

Many forms of cooperation exist within natural systems, such as mutualism (where both parties benefit) and commensalism (where only one party benefits, but the other is neither harmed nor receives any benefit). An example of mutualism is coral and zooxanthellae algae. The coral polyps provide the zooxanthellae with a protected environment and compounds they need for photosynthesis. In return, the zooxanthellae produce oxygen and help the coral to remove waste.

There are also examples in nature where we can see both cooperation and competition happening at the same time. For example, many frog and toad species will gather together en masse for mating when the seasonal conditions are right. Males are competing for a female partner and show signs of call variation depending on the level of competition. It is also thought that these mass gatherings limit their vulnerability to predators, who can't work out where an individual frog or toad is, because the croaking is coming from many places at once. A case of safety in numbers, which is also the case for the development of human habitations through the cooperative and protecting nature of clan, tribe and village.

We can clearly see both cooperation and competition in natural systems as effective strategies, so why is this principle so focused on cooperation as the best option? Surely applying both at appropriate times would be an example of ***working with nature***?

The ethical influence

When we consider our permaculture ethics, especially ***people care*** and ***fair shares***, cooperation is clearly the winner. The competitive nature of people has become dominant in our cultures and economic systems, consistently being a key factor towards destroying the earth, breaking up communities and the unequal sharing of resources. When competition is combined with ***efficiency*** then we may have highly effective solutions but normally at a terrible cost as the powers of competition are ramped up. As we lean into the permaculture ethics we are steered away from competition and yet the pressure on us from society leads us towards a competitive approach as a means of survival. Many permaculture designers unlearn or limit their competitive natures and explore ways of being more cooperative with others and with nature. But maybe there is still a place for a little competition within the ethical framework.

Our competitive nature

Humans have been dominating the landscape for many centuries now with housing, farming and industry. We have out-competed many other animals, plants and each other to take control of space and resources, in many cases resulting in the major loss of habitat and species, alongside prejudice, racism and colonialism. The competitive righteous way is horrendous and has the tendency to become very, very unpleasant, at the very least leading to selfishness and cheating.[11] Once that dominant competitive attitude is in place then cooperation and ***diversity*** tend to be suppressed.

In our farming practices it is commonplace to cull and restrict many species of competing animals, such as crows, gulls, rabbits, pigeons, rats and a whole range of insects. They all want the food that we grow too. Even in an organic or permaculture farm setting you would not be cooperating with a rat about food crops. Many of those larger 'pest' animals, such as rabbits and pigeons, taste nice, which could be an example of the ***problem is the solution***, but not if you are vegetarian/vegan. Inevitably the food crops we find tasty are also tasty to many other animals, and in growing such foods we create a competitive situation with those that live alongside us. In some cases that might be seen as a ***fair share*** to the local wildlife, such as hedgerow birds eating some soft fruit. In other situations we have shot ourselves in the foot by removing predators, such as the fox, who would otherwise control some of the 'pest' animals.

I am frequently asked how to manage rats on allotments and smallholdings. Being in competition with these animals does not mean we have to hate them. They are just being themselves, trying their hardest to survive and doing what they know best. At the same time, if you are growing food, managing compost systems, or keeping livestock, then I highly recommend having a rat management strategy, however that fits with your ethics. We also have lots to learn from them, especially when it comes to ***resilience*** and ***adaptation***.

Our competitive nature can impact our mental wellbeing as well. We have a tendency to compare ourselves with others resulting in superiority and inferiority. I am better than you at this, you are better than me at that. This can be useful to allocate roles and responsibilities but can also lead to some disturbing competitive behaviour. According to the nonviolent communication approach, developed by Marshall Rosenberg, comparing ourselves to others is a form of violence.[12] When our leaders generate circumstances of insecurity and anxiety through an apparent scarcity of resources then we have a choice to compete with our fellows or we can choose to cooperate with each other.

Challenging edge and striving together

Is there anything wrong with a little healthy level of competition? Are we rejecting a valuable aspect of nature that could help us? A level of competition, can be argued, is essential for our survival. David Holmgren outlines the benefits of competition:

> *Competition in nature helps test the vigour and fitness of individual organisms, or a species, for particular conditions.*
>
> **David Holmgren**[13]

We need challenges and risks in our lives, and sometimes we don't have a choice in that, such as the transition from teenager to adult. A competition with yourself to better yourself can be healthy. These ***edges*** create excitement, anxiety, momentum and drive. They encourage us and progress us towards learning and adventure. No doubt we have all found ourselves in a competitive situation, maybe that job application, sports event or a board game. These can be seen as win/lose situations but when we lose we have the opportunity to turn that ***problem into a solution*** through learning. When you have lost the game then ask how you can improve and what you can learn from the situation to decrease the probability of 'losing' next time. Focus on being the best that you can be rather than comparison and winning. This is a healthy form of competition.

The word compete comes from the Latin *competere* meaning *to strive in common.* The use of the word in the commercial sense of *striving for command of a market* came much later in the 1800s. Is there a place for returning to an older meaning of competition as *striving for something in common*? If we can understand the nature of excessive competition then we may be better placed to avoid its traps and find its positive side.

Permaculture designers do strive in common for something big. We are engaged in a competitive battle alongside many other green allies. We do want to win a battle of ideas. We are in competition with the modern aggressive capitalist model. We need our ethical approach to life to win because we are convinced that it is one of the keys to saving planet earth and building healthy communities. To do that we need to cooperate and compete.

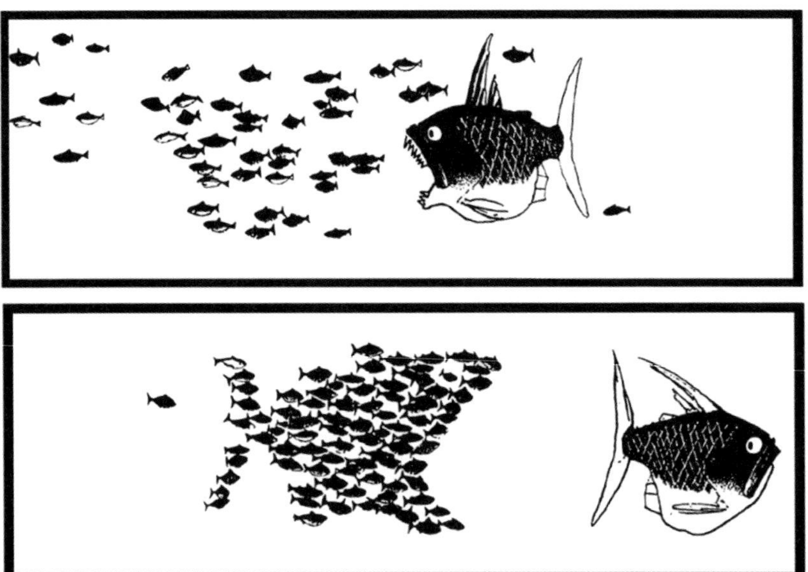

'Big fish, little fish' from a Radical Routes pamphlet © Radical Routes

The power of cooperation

Cooperation has been a powerful principle for me. I have been involved in many cooperative adventures including various community groups and cooperative businesses, including Abundant Earth.[14] The modern cooperative movement was largely started by the Rochdale Pioneers in 1844,[15] who developed their own set of cooperative principles, such as *member economic participation* and *cooperation amongst cooperatives.* I was deeply inspired by cartoon images representing the power of cooperation in the 1980s and 1990s.

Radical Routes[16] are a coop of coops established to enable grassroots control and economic revolution through cooperation. They have been particularly focused on the development of housing coops, which enable people that live in a house to co-own the house and thus are not paying a private landlord driven by profit. Abundant Earth is a member of Coops UK,[17] a network of thousands of independent coops. Coops UK frequently provide us with details of the economic strength of cooperative businesses, such as how during 2022 the number of coops increased, compared to the number of businesses overall which dropped in that same period.[18]

We have some ancient sayings about the power of cooperation, such as 'the whole being greater than the sum of the parts' and 'many hands make light work'. Holmgren added these phrases into the development of his principle ***integrate rather than segregate.***[19] Permaculture encourages self-reliance, which means seeing what you can produce, cooperating with those in your community to trade or exchange your surplus, unlike self-sufficiency which places an expectation that we can do it all on our own. Our society still largely encourages individualism and self-sufficiency. If you find yourself saying to yourself, I can do this on my own and I don't need any help, then you are resisting the cooperative spirit.

Maybe next time you are in a strange town and need directions, ask a local and don't just rely on digital maps. We have to remind ourselves that a task done with others is often easier and more fun.

Cooperation is great. We use it as a vehicle to work together, to collaborate, form teams and to achieve bigger things than we could on our own. There is evidence now that teams of people trained to cooperate with each other are more productive than teams with an internal competitive attitude. So encourage your favourite football team to work out their internal social dynamics and they might perform better.[20]

The art of balancing cooperation and competition

Cooperation is not entirely blame free and holy! Businesses in the same sector will frequently cooperate with each other to fix the price of a product or service. Recent research exposed Uber drivers cooperating with each other to actively fix higher prices. Maybe there are times when you are too cooperative with those around you to avoid conflict, allowing them to make decisions for you and influence you too much. We need to apply this principle with the permaculture ethics and other principles such as ***integration***.

As a veg box scheme manager it would be easy to assume that I am in competition with other neighbouring veg box schemes. There are moments when I discover that a veg box scheme based over 30 miles away is supplying someone only a couple of miles from our smallholding. Some national veg box schemes are highly competitive and appear not to care about the very local schemes like ours. We actively cooperate with other food growers in our region: they might be wholesalers to us at times, advisers and supporters in other moments and even regional project collaborators. There is a lot to learn from each other even if we are at times ever so slightly competing with each other.

Sports teams can have the paradox of cooperation and competition existing simultaneously, with cooperation internally within the team, and competition externally with other teams. Both need to exist in these systems otherwise the fun and challenge would be limited. Research shows that the smaller the teams the greater the external competitive rivalry, with the greatest being matches between two individuals.[21] Just think of boxing matches if you have any doubts about this. So if you have a tendency to be drawn into competitive challenges and find your ego getting out of hand, then getting involved in larger team sports where internal cooperation is encouraged will be healthier for you.

	Cooperation	**Competition**
Plus	• Builds relationships • Can go further and last longer • More resilience • Better for the group	• Encourages growth • Can stimulate motivation • Opportunity for learning when you lose
Minus	• Restricts personal freedom • Requires facilitation • Takes longer • Requires personal compromise	• Not everyone thrives in competition • Some people have had experiences of never winning or never feeling good enough • Can lead to powerful destructive forces
Interesting	• Learning to cooperate well can lead to personal and social growth	• What would some of our designs look like if we included an ethical quantity of competition?

Example of using the Plus, Minus and Interesting (PMI) tool devised by Edward de Bono

Masanobu Fukuoka, the author of *One Straw Revolution,* utilised the competitive power of cover crops, such as clover, to out-compete the weeds around the main crop. The timing of the sowing of the cover crop could be very important to maximise its competitive powers i.e. sow the cover crop before the weeds have become established. We can also be selective predators when we thin our seedlings out, inevitably choosing to keep the ones that we think look best.

We can see that both cooperation and competition have pros and cons. So let's use them both in balance where appropriate and ensure that they are used with the permaculture ethics. Let's move away from the simple soundbite of *cooperation, not competition.* It's more complicated than that.

Diversity

Written in collaboration with Chris Evans and Katie Shepherd

Diversity is normal

Diversity is everywhere, everything is unique. We are all different and there are loads of things. We can't deny that diversity is normal. Diversity comes in many forms including biological, climatic, geographical, or relating to ethnicity, gender, age, wealth, religion, sexuality, abilities, approaches and culture.

Take a spoon of topsoil and know that it is estimated to contain a billion microscopic cells and possibly 10,000 different species.[1] In our guts there are now estimated to be up to 1000 different microbial species, mostly helping us with digestion.[2] Our planet is full of a massive biodiversity of species, millions of them, including over a million species of insects alone. Around the planet that diversity varies as climate and geography vary, from extreme heat to extreme cold locations, from dry to wet, from flat to mountainous, from savannah to forest, from tiny streams to deep oceans, with each species occupying its own niche, their preferred habitat. It's all absolutely amazing!

As humans we have all sorts of shapes, sizes, races, beliefs, abilities, opinions, genders, locations, languages and cultures. On top of that we can add the constant

Diversity in a teaspoon of topsoil and mulch © Wilf Richards

changing and dynamic impermanence of the world. It's mind-blowingly complex. It is no wonder we crave simplicity, stability, predictability and control. It seems like absolute chaos out there but it all needs celebrating and embracing. Tragically due to power imbalances, prejudice, privilege, ignorance and fear, diversity has been suppressed in many ways across our societies. As permaculture designers we have our ethical foundation, which includes ***people care*** and ***fair shares***, highlighting the importance of equality, equity and challenging unhealthy power dynamics.

Diversity is a core principle

The early permaculture books referred to diversity a lot, especially in relation to our understanding of natural systems and how ecological diversity supports the stability and hence ***resilience*** of those systems. It's there right at the start in the first book *Permaculture One, A Perennial Agriculture for Human Settlements* by Bill Mollison and David Holmgren published in 1978.[3] John Quinney wrote about diversity as one of his principles in his 1984 *Mother Earth News* article.[4] He focused on plant diversity and how we favour that over a monoculture any day but he added in a caution about how plants also compete between themselves for light and nutrients so too much plant diversity can be detrimental. He describes the importance of beneficial relationships as just as important as diversity, and so in some situations we need to limit the amount of diversity to increase productivity.

In 1991, we see diversity clearly named as a principle in the *Introduction to Permaculture* book. Here the principle is worded as ***diversity, including guilds***.[5] They continue the exploration and benefits of diverse plant polycultures and dense planting of perennial crops, in contrast to the modern straight line monocultures. The importance of diversity features in just about every permaculture book, whether that be increasing the number of ***elements*** in a system, having a range of strategies, genetic diversity, biological diversity and cultural diversity. It is clearly core to our understanding of the world and an essential ecological principle for us to apply.

This principle is called ***use and value diversity*** in Holmgren's book,[6] in which he reminds us to *not put all of our eggs in one basket*. This proverb, probably originally from Miguel Cervantes's book *Don Quixote*,[7] tells us to not invest all of our resources and energy into one system. Instead we need to hedge our bets and diversify our efforts into many options otherwise we risk the collapse of our setup. Holmgren links this principle to ***succession***, as diversity over time, and to ***stacking*** or layering, as diversity through space. He describes how permaculture ***integrates*** biodiversity into its landscapes rather than segregating it and how species diversity can reduce competitive relationships as each species type focuses on its niche of different needs and resources. He shares evidence that diversity is a proven survival strategy for plant and animal populations, and how there are various factors of diversity that contribute to ecosystem stability.

Patrick Whitefield's book *The Earth Care Manual: A Permaculture Handbook for Britain and Other Temperate Climates* (2004),[8] includes a lengthy section about the

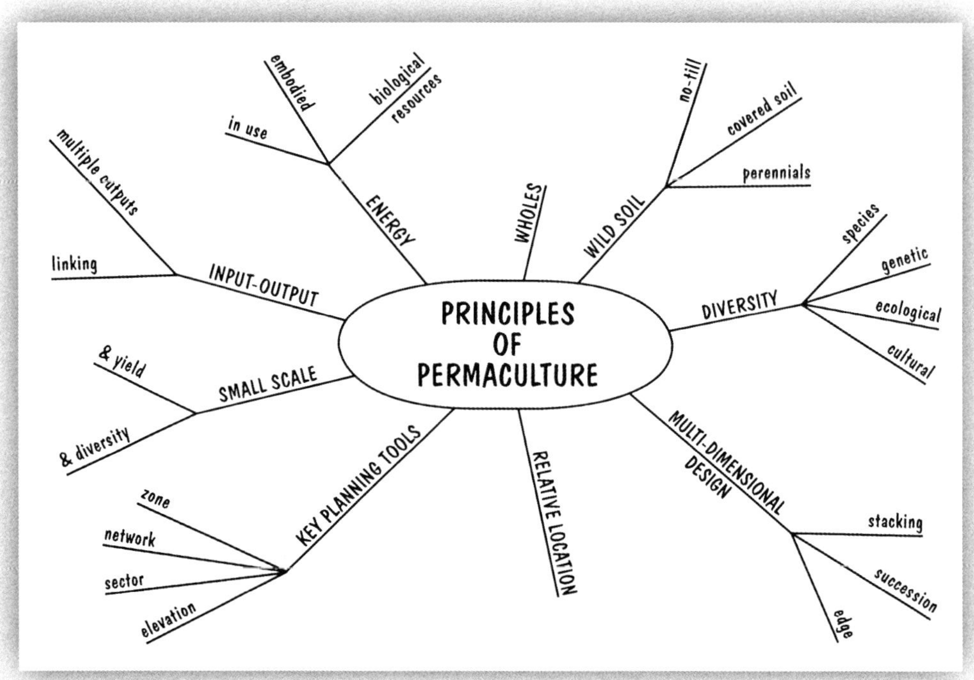

Patrick Whitefield's map of principles from his The Earth Care Manual

importance of diversity. Here he explores the relationship between diversity and niche; for example, ash trees and wild garlic occupying the same location but avoiding competition by having different timing of leaf growth. We can apply this principle of diversity and niche in livestock management by, for example, having sheep in with cows as they have different grass length preferences.

Whitefield also elaborates on the connection between diversity and stability. More diverse systems are more stable because there are a greater number of species to carry out key functions of the whole ecosystem. This is known as *functional redundancy*[9] and makes diverse systems more ***resilient*** to system shocks. As one species is lost, another one may fulfil the functional gap, even if not as effectively. Our decline in biodiversity results in an ever increasing risk of ecosystem collapse because we are heading towards the last key species to maintain a key ecosystem role. Eventually the redundancy in the system will be lost if we continue with the current rate of biodiversity loss. Many habitats are showing signs of hanging in there, even giving us the illusion, or at least to the untrained eye, that everything is OK. Patrick's concern, back in 2004, was that the hypothesis of species redundancy would mean we can get away with a big loss of biodiversity before the effects are noticed. Looks like he was right. Evidence has grown amongst the scientific community, supporting John Quinney's original writing in 1984, that it is not the diversity of species that creates stability but the diversity of functions and relationships.

Our bad habits that crush diversity

We need to recognise the massively negative impact we have upon diversity in all its forms. We can crush diversity all too easily in our industrial capitalist society. Whether that be in the loss of crop diversity, destruction of habitats, our intensive consumer habits or our prejudice that excludes groups of people. All of these bad habits clash with the permaculture ethics.

Our urge to organise and simplify the world around us can help us make sense of diversity and to assist in communicating about our experiences. But this step towards simplicity may lead to generalisations and predictability. We are creatures of routine and we like to know what's coming next. When you buy a delicious item of food and you want it again, you want to know it will be the same glorious experience as before. This certainty of factory-line products and our consumerist society has contributed to the decline of diversity. High streets and housing estates in every town can look the same. Our countryside is full of monoculture: giant fields filled with one crop, lacking diversity, all in the name of ***efficiency*** and control.

The biodiversity of the planet and diversity of cultures is now seriously threatened. We are witnessing and contributing to a major extinction wave. We can see that humans are almost entirely responsible for this ecological collapse by simply studying places where we have been excluded and wildlife has returned, such as Chernobyl.[10] We can be stewards of this earth and enable biodiversity to thrive. We can also accept our differences and enable a diversity of cultures to thrive.

Can we have too much diversity?

We have already explored the importance of beneficial relationships over diversity. If all of the plants we put into our gardens were completely different species then the opportunities to cross pollinate would not be available. Some plants, including most types of apples, need fellow plants to set fruit, as many cannot self-pollinate.[11] If we only had one type of each plant and one of them died then we would have lost that species.

I once visited a forest garden that had been very well stocked with plants in its early days and then the owner became ill and the plot was unmanaged for a long time. Some of the plants could grow faster and taller, and so started to dominate the site. Meanwhile others, more delicate, had been shaded out, oppressed and were suffering. Every garden needs maintenance for those checks and balances to maintain diversity. There is also a pattern amongst permaculture gardeners to try many ideas in the early stages of a land project, applying a high degree of diversity, but with time, as lessons are learnt and some crops fail and others thrive, we can have a tendency to simplify our project and concentrate on what works best. This demonstrates a balance between the principle of diversity and that of ***efficiency***. Holmgren refers to this as culling of early experiments to make things more productive,[12] and as a natural pattern that is commonplace.

Technology has recently had a problem with too much diversity of connectors and plug types. How many times do we need to change that connector for our mobile phones? Fortunately there is a move now towards some standardisation of phone connectors but someone may invent a more ***efficient*** new connector and so we may need to change it all again.[13]

We all want to celebrate the diversity between people but there are moments where differences can be a problem. It is great to have something in common with your fellows even if it is just a common language to communicate and share experiences. For anyone that works in a group, you will recognise those moments when everyone seems to have a different opinion and it is hard to identify the consensus and best way forward. In some cases, where agreement can't be reached then conflict may arise. So finding ways to achieve common ground and resolve conflicts is very important. This could be done through facilitation, mediation or using thinking tools that encourage a diversity of opinions, such as Edward de Bono's Six Thinking Hats.[14]

Diversity and inclusion

The diversity of opinions, approaches, skills, experience and knowledge amongst members of groups is to be celebrated. We need to remain open and curious to someone else's viewpoint, and celebrate that they are different from ourselves. Even in a small group the diversity of culture may be minimal but diversity of opinion can still be great. Awareness of, and celebration of, neurodiversity has grown a lot in recent years, as once 'strange' viewpoints turn out to be the best way forward. We may need to apply positive bias and techniques to actually include someone in a group. Diversity and inclusion strategies are different and both are needed. Diversity advocate Verna Myers describes the difference as "diversity is being invited to the party, inclusion is being asked to dance".[15]

In late 2022 and into 2023 Katie Shepherd and Kit Dickinson facilitated a programme of events for permaculture practitioners with disabilities and/or chronic illness, which resulted in the publication of a zine.[16] During lockdown, online events were the only option for all of us and this gave better access to those with various disabilities such as chronic fatigue. The zine shines a light on the struggles for those with such disabilities, and the importance of maintaining those inclusion strategies post-Covid.

The explorations and stories included are focused around the Holmgren principles with some great examples, such as the need for extra rest as a form of ***catch and store energy***. It brought to my attention the need for special considerations and the struggles of inclusion, the ***limiting factors*** that exist in others' lives that I do not experience myself and the need for non-judgemental compassion. We need to remember that everyone is just trying to do the best they can with the resources they have available, and in some cases those resources are very limited and extra effort needs to be made by those with privilege to include those without. Another

Diversity and inclusion in a community garden © Katie Shepherd

example is the ongoing challenge for those in wheelchairs accessing buildings. Despite legislation in the UK to encourage accessible buildings there is still the option for developers to choose the cheapest options with limited accessibility.[17]

As well as increasing inclusion and accessibility we also need to break out of our routines, step out of our comfort zones, push towards the ***edge*** and challenge our assumptions to find a different way of being. We need diversity in our thinking, which in turn can lead to new behaviours. Let's get unstuck and try to be more fluid. Let's get creative and experiment with materials. Let's make something new and add to the massive quantity of diversity. Let's simply allow the natural tendency towards diversity to be unhindered and support our sanity. Doing the same things over and over again sure is boring. We need diversity of work, projects, people and experiences. Even when we are stuck in a rut and experience the mundane, we need to switch on our nuance detectors and understand the subtle diversities. After all, every moment is unique in this impermanent world and everybody needs to be cherished.

Enabling diversity to thrive

A simple place to start with diversity is at home in what you eat. Tim Spector,[18] the genetic epidemiologist, who wrote numerous books on gut health and was a lead researcher in that field, concluded that we need over thirty different plant species in our diets every week. This diversity of plants feeds the various microbes in our guts preventing any one microbe from dominating. If you only eat one type of food then you will only be feeding a few types of microbes in your gut. When you include grains, seeds, vegetables, fruits, nuts, legumes, herbs and spices in your diet then achieving thirty plants a week becomes surprisingly easy.

In 2020 I carried out a gut health design and one of the changes I made was to switch from about four plants in my breakfast to over ten every morning. The cereal mix had become a DIY muesli 'trifle'. It tastes amazing and my guts feel fantastic too. By adding probiotics to our diet too, such as kefir, sauerkraut, kombucha and natural yogurts, we can fertilise our good gut microbes even further. By using a wider diversity of plants you will also be starting to influence the diversity of crops grown in the landscape around you. The majority of our calories and proteins from plants are derived from just three species, the top one being wheat.[19]

Our next step towards diversity is the food we grow. We need to shift towards a multitude of small polyculture farms using a diverse range of farming methods, such as agroforestry, regenerative farming, biodynamics and organic farming. This will encourage biodiversity to thrive either through direct regenerative practices, adding wildlife ***edges***, utilising rare breeds of livestock or by banning the use of chemicals, like pesticides, that harm wildlife. We can even shift our personal monocultural allotments to diverse polycultures of mixed planting, maybe as ***stacked*** forest gardens. There is always room for more plants, after all our ***yields are only limited by our imaginations*** and there is always one more niche that can be filled, typically at an ***edge*** of the space. Polycultures can have a higher overall yield, are more likely to be ***resilient*** to pests and diseases due to the increased diversity, provide a broader range of produce and if scaled up then a wider range of potential incomes.

Adding more elements to any system will increase the diversity to a certain point. But, as already described, it is not only the quantity of ***elements*** that we add into a system that counts but also the quality of relationships and ***functions*** between those ***elements***. For example, some plants ***cooperate*** well with other plants, whilst others can be antagonistic to each other. We can deliberately put together a grouping of diverse plants that will support each other and minimise the potential antagonism. These are called 'plant guilds' and they take advantage of our knowledge of niches. These guilds aim to minimise competition between the plants through utilising resources at differing times in the seasons, providing shelter or protection to each other, assisting in pest control and even an exchange of nutrients.

In terms of biodiversity we need to apply ***earth care*** and ***fair shares*** ethics by observing the landscape carefully, limiting negative human activity and allowing natural systems to regenerate. We could, for example, switch our perspective of weeds, by using the principle of the ***problem is the solution*** and recognise the various positive roles of weeds as soil regenerators, wild foods, medicinal plants and even a potential income source. In the UK the highly invasive plant Himalayan balsam provides nectar for bees and edible seeds for us. This is not a plant that needs encouragement and is illegal to plant in the UK but equally our fight against it could have a different approach. We need to ***integrate*** nature into our lives and give it a voice and rights. We also need to apply ***people care*** and ***fair shares*** ethics together to support the oppressed and marginalised and ensure the redistribution of power in our human societies. There is a lack of diversity at the top of the ladder but let's not just change that but let's get rid of the ladder too. The top-down hierarchy dominant power structures trample diversity in so many ways. We need better ways of making decisions, such as sociocracy,[20] that enables all voices to be heard. We need to suppress those overpowering dominant factors and take the power back.

Edge Optimised

Written in collaboration with Ed Tyler and Ruth Starr-Keddle

The ecology of ecotones

For us to understand this principle about edge, we first need to understand more about its origins from ecologists. Ecologists refer to the overlapping area of two ecosystems as *ecotones*.[1] The term ecotone comes from the word *ecology* and the Greek word for tension *tonos*. It was most likely coined by the 'father of biogeography' and co-founder of the theory of natural selection, Alfred Russel Wallace,[2] in the middle of the 19th century. He was a prolific explorer, scientific researcher and pioneering environmentalist. Ecotone is defined as the transition area between two biological communities. This transition area could be narrow or wide, local or regional, a gradual change or distinctly visible. Examples of ecotones can include where the water and land meet (for example seashore, riverbank, flood plain and wetland), where two types of water meet (river meets the sea, such as estuary and delta) or where two types of landscape meet (woodland edge, woodland clearing and hedgerow).

The characteristics of ecotones are that they have more resources, species richness and increased biodiversity, and this is called ***edge effect***[3] by ecologists. One of the richest ecotones are the coral reefs, evident by only occupying 0.1% of the world's ocean area but providing a home for at least 25% of all marine species.[4] There can be a wide range of microclimates and niches in ecotones, for example, you could find both light-loving and shade-tolerant plants, or both woodland and grassland plants within the same ecotone. These characteristics are because there are species that specialise in just one of the habitats and then there are species that can occupy both habitats. On a riverbank we will have the plants that need drier conditions further away from the bank, through to plants that can tolerate both dry and wet conditions, like willow and alder, and then plants that need wet conditions all of the time. The same ***pattern*** occurs amongst animal species, where some specialise and some can cross over the two habitats. For example, fish will stay in the water whereas amphibians and otters can occupy water and land, and most mammals will stay on the dry land, only to venture to the water for drinking.

As we step from one biological community to another, across the ecotone, then we may see clear physical sharp differences, or the change might be gradual as two habitats merge, or they may only be visible through measuring equipment such as pH testing of soil or salinity levels in rock pools. If you ever walk out of a woodland and into a pasture, you may be struck by the big temperature difference, which could depend on the season and strength of sunlight. I am still often surprised by how cool it is in the shady woodland on a hot summer day and the jump in temperature when I head up to the sunny market garden. Ecotones are influenced by geography, geology, climate and biology, all acting together.

An ecotone could also be a physical boundary preventing some species from moving any further, such as bodies of water or mountain ranges. This boundary effect can be key to the evolution of species, as a ***limiting factor*** of movement is imposed, causing species to ***adapt*** to their local conditions. One major example is the Wallace Line[5] in Indonesia, named after Alfred Wallace, where there is a big difference in species on either side of the line due to deep water separating the land masses and islands.

Woodland, scrub and pasture edges at Abundant Earth © Wilf Richards

Permaculture gets an edge

Bill Mollison and David Holmgren wrote about ***edge effects*** in the very first permaculture book[6] in 1978. Mollison and Holmgren share their observations, research and the productive possibilities about edge habitats. They go into quite some detail about the effect that edge has in and around forests and the influence on microclimates, especially temperature and wind variations. The concept of working with edge, and normally with the intention of ***increasing edge*** has inspired the development of polyculture systems, wavy edged ponds and keyhole beds.

The term ***edge effects*** was formally added as a principle in Mollison and Slay's 1991 book.[7] They describe edges at all scales, from soil particles touching water, to where frost pockets meet frost-free areas and all natural or artificial boundaries.

They also highlight the importance of productive edge habitats for human settlements. Consider villages placed where woodland meets river, or the traditional dwellings at the edge of islands with plenty of opportunities for food out to sea and inland too, providing a rich and varied diet. Mollison and Slay recommend dwelling where at least two productive ecosystems meet or changing the landscape to increase those edges and thus increase productivity. They also recommend using traps to catch energies and nutrients entering a site from beyond the edge and prevent them from leaving, for example leaves being blown up to a fence. This was before the term ***catch and store*** was being used but this is exactly what is being described. Finally, they explore how through our understanding of ***patterns*** we can increase edges in our designs, including the classic herb spiral and creating productive lobe shaped clearings in scrub and bramble areas. Mollison was inspired by many Indigenous and sustainable food cropping systems where edge is clearly a key part of the design, including the chinampas[8] of Meso-America, and other floating gardens. Angelo Eliades, author of the Deep Green Permaculture website,[9] explores the relationship between ***patterns*** and edge effect further on his website.

Edge appears consistently as a principle in just about every permaculture book since. Patrick Whitefield in *The Earth Care Manual*[10] put edge together with ***stacking*** and ***succession*** as the three principles for multidimensional design. Rosemary Morrow describes it as ***valuing edges and marginal*** in the second edition of *Earth User's Guide to Permaculture* (2006)[11] and Toby Hemenway[12] asks us to ***optimise edge***, by increasing it or decreasing it as appropriate. It is typically referred to as a *principle of ecology*, although Colleen Stevenson[13] and Sarah Spencer[14] both refer to it as ***use your edge*** with an implication that it is just as much about our attitude.

In his book, David Holmgren[15] expanded on *edge effects* and named this principle ***use edges and value the marginal.*** He adds the proverb 'don't think you are on the right track just because it is a well-beaten path', encouraging us to explore wild and unknown territories, also indicating an attitudinal aspect to this principle. Holmgren has helped to expand this principle into more than just land-based applications, as can be seen on his website,[16] which includes examples such as a library at a bus stop and the twisty shapes of pasta for carrying the sauce. He includes a wide range of maximised edge examples from the landscape scale-down to the vast surface area in our lungs and the connections between soil and the elaborate roots of plants. He shares details of how Indigenous Aboriginal use of fire to manage the landscape creates plenty of edge through the creation of more ***diverse*** habitats. He also explores urban examples of edge too, such as the suburbs (edge between town and country) and the shop front (edge between shop and street). Holmgren also provides a positive exploration of marginal aspects of our society, looking at the joys of wild food and opportunities in rundown neighbourhoods.

Increasing edges for habitats

Conservationists use their knowledge of *edge effects* to support biodiversity and ensure better ***earth care***. They are constantly trying to make sure that the messy edges are left alone or encouraged. They refer to this as *environmental heterogeneity*, in other words creating a variety of conditions. In pond construction different depths will be included to provide a range of habitat zones within the pond, some suitable for deep-water species, some for shallow-water species.

Wading birds in the UK desperately need conservation support. One way in which they are given this support is through the introduction of scrapes[17] in wet upland sites. These are shallow dug out sections of the landscape to expose water to the surface. They are best done as wiggly lines to maximise edge. These long thin ponds provide increased habitats for invertebrates which in turn provide feeding opportunities for amphibians, birds and mammals. This is of particular importance for the chicks of wading birds who will feed at these locations.

Other edge related conservation efforts can include roadside verge management[18] for wildflowers, the introduction of hedgerows, leaving arable field margins uncut for pollinators and riparian buffers.[19] Riparian buffers, which are the vegetation edges along riverbanks, are particularly important for biodiversity, preventing soil erosion, acting as biofilters for water entering the river and maintaining a stable river water temperature. The best riparian buffers are diverse, wide and include trees.

Are there times when we need to minimise edges?

Increasing the edge is not always the best strategy. For example, when we need protection, minimising the edge could be useful. This could be as a safety measure such as guardrails or protective barriers on flat roofs or when working at heights on a building site. It might be a fence near to cliff edges where falling is a possibility. It may also apply to sharp corners on objects that could be softened or putting knives into a protective case. I am also reminded of the ancient Scottish and Irish structures referred to as crannogs.[20] These are dwellings on projections into a loch, typically made of timber or stone that provided only one entrance onto the site, typically a narrow passageway that could easily be defended. Thus edge with the water is maximised for food but edge with the land is minimised for protection.

Minimising edge also applies to protection against disease. Certain diseases are transmitted by contact and in those cases minimising the edge of contact is desirable. We can keep our distance and avoid unnecessarily spreading the illness to someone who may be more vulnerable than ourselves. Some diseases, such as the parasite *Leucochloridium*[21] which infects freshwater snails, triggers behaviour changes that make the snail move into more exposed places where it is more likely to be eaten and pass the parasite onto the next host. Such parasites and most viruses thrive on crossing the edge of immune system barriers.

Where we want ***efficiency*** and speed, we need to minimise the edges too. This could be between two machinery parts or a streamline design. All machines have moving parts and where two parts touch or interact then oil can be used to lubricate, minimising the friction and increasing the ease of movement. Alternatively polishing a surface, as we experience on a slide, can minimise micro bumps and increase the speed. A streamline shape, for example in certain fish, with scales all going in the same direction, enables quick movement to evade predators or inversely to catch prey. Equally we may want to increase the edge to add friction, such as on a car tyre or shoe tread.

We also need to minimise the edge where we want to minimise heat loss. A house with lots of wiggly edges has more surface area, all of which increase the likelihood of a faster heat loss. Animals with a high surface area to volume ratio lose heat at a slower rate, which is why animals in colder environments have evolved to be bigger and also why elephants have such large ears as a means to remove excess heat. We can keep a space, for example a house, the same size in terms of area but it can have a wide variety of edge lengths depending on the shape. This could increase microclimatic opportunities in a garden so in that scenario we probably want a shape with lots of wiggly edges, but in a house the same wiggles would increase heat-loss potential.

Optimising edge for productivity

Permaculture designers are keen on increasing productivity of landscapes in balance with biodiversity benefits, and using the edge principle can help to a certain degree. Maybe you have an under-utilised space at the edge of your garden that can become a productive area. Maybe we can consider the ***relative location*** of two crops and if there is a known beneficial connection between them, then we can bring them closer together: for example, adding French marigolds to the tomato crop to dispel whitefly.[22] We can add in wiggly pathways, keyhole beds and multi-dimensional planting or we could increase the number and ***diversity*** of plants, shifting towards a polyculture.

These strategies can thrive at the small scale of the private garden. As the scale increases though, the edge strategies need to change. Wiggly paths and curvy beds are not practical on a field-scale operation of veg harvesting where we need some straight lines for ease of picking and using machinery. At Abundant Earth our market garden is only one acre, which is small but even there we need most of the beds to be straight for ***efficiency***. We still have plenty of edge strategies in place by breaking the space up with a couple of internal productive hedges and the edges of the garden have more perennial crops. At an even larger field scale of operation the need for straight lines increases and instead strategies, such as alley cropping and agroforestry methods, can add edge to the system. On all scales we can also increase the seasonal edges by adding in polytunnels and greenhouses, which extend the length of time for growing crops.

Alley cropping at Can Lliure © Tobias Sonrisa Sturmer

Value the marginal

Holmgren expressed this principle as ***using the edge and valuing the marginal.*** The marginal might include the areas of land that are least productive with a lower economic return but possibly higher in biodiversity. They may also be the areas that provide mushroom foraging opportunities, wild seasonal foods and medicinal plants; places where observation of nature can excel or seasonal marginal grazing can take place. These places need protecting. We do not need all of the landscape to be economically productive. The intensification of farming can strip out these marginal edges with, for example, the removal of hedgerows, scrub or wildflower edges. All in the name of progress and ***efficiency***.

The marginal could also be the edges of society, places of poverty and limited resources. We must value these places too by investing in them. The addition of the ethic of ***fair shares*** is key here to support a better distribution of resources. We need to ensure the hard edges between unequal communities are dissolved and enable a more equitable distribution of resources to take place. The number of people living at those edges of our societies are an indicator of the levels of equity and equality. The gap between the richest and the poorest has continued to grow for a long time now. It is one thing to see opportunities in marginalised situations but it is another to see people stuck in the margins unable to get a leg up and find comfort. The gap needs to stop widening and start narrowing.

Out of your comfort zone

Have you ever 'felt on edge'? When we move outside of our comfort zone we may be going beyond the boundaries of our skill set into an area that is too challenging for us and that could trigger anxious feelings. If we can increase our skills and capacity or decrease the scale of the challenge then we can stay in our comfort zone. By tackling a new situation one tiny step at a time, we may push back the edges of our comfort zone. This is related to flow state theory[23] and is explored further in the ***small and slow*** chapter.

One strategy to push yourself outside of your comfort zone is to use a variation of the 80:20 rule. In this context we could use it to spend 80% of our time in a comfortable place, and push ourselves 20% of the time.[24] This is a successful and commonly used strategy by athletes to push their performance barriers. For example, if you go running five times a week then four of those runs could be at a comfortable pace and distance or if each of your runs is for 40 minutes then push yourself for only eight of those minutes.

Another comfort zone that we can push is our social media bubbles. It is all too easy to seek people with similarities, providing confirmation and reassurance of our own beliefs but this comes at a price. That bubble could result in limiting belief systems and being trapped in the certainty that your way is right and best. We need to encourage ourselves to engage with difference, being open and curious about others who are not the same as ourselves.

We can look out for creative collaboration opportunities to go beyond the edges of our beliefs: different people brought together, each one with their own skills, knowledge, experience, strengths and weaknesses, ***cooperating*** with each other. Together we can do so much more as the increased capacity can push back edges and move mountains.

Energy Efficiency

Written in collaboration with Cathrine Dolleris, Sally Hughes and James Taylor

Life, universe and everything

Having a chapter including the word energy could just simply be too big so let's try to focus. Understanding and working with energy in all of its forms is useful to being human or any form of life, and permaculture has had a few things to say about it. There are plenty of scientific laws about energy, such as those about thermodynamics[1] and conservation. They are complicated but you might recall simple explanations, such as *energy can be transformed from one form to another, but can neither be created nor destroyed.* As permaculture designers we are interested in a number of aspects about energy, such as what form it takes, where it comes from, where it is going, how I can use less of it, how I can ***obtain a yield*** from it, and whether there are pollution or ***waste*** issues as a consequence of using a form of energy. We are interested in food, fuel, fibres and shelter and these all require energy and are forms of energy.

Right from day one of permaculture we can see the need to understand energy. *Permaculture One*[2] refers to the shift from energy-expensive agribusiness to a low-energy high-yielding agriculture. A key motivation has been around the running out of finite fossil fuels and the need for alternative ***renewable*** forms of energy. And so permaculture has developed and utilised tools and strategies to conserve energy, ***catch and store energy***, map inputs, outputs and flows, reduce energy consumption, switch to greener energies and be more efficient with energy. As Bill Mollison said in his *Designer's Manual*, 'Permaculture as a design system contains nothing new. It arranges what was always there in a different way, so that it works to conserve energy or to generate more energy than it consumes.'[3] Despite a strong focus on energy it was a while before any principles appeared using the word, although we could see hints of it in ***least change*** and explorations of it around the ***yield*** principle. The main energy focus in the early years of permaculture was around placement through design. This manifested as tools, such as zones and sectors, alongside the principles of ***relative location*** and ***efficient energy planning***, which both appeared in the 1991 *Introduction to Permaculture*[4] book. Despite the naming of that principle as ***efficient energy planning***, which includes the word efficient, I have put the details about that principle into the ***relative location*** chapter. This is partially due to the article back in 1984 by John Quinney.[5]

I haven't put everything there is to say about energy into this single chapter. That would be mad. You can see aspects of energy in many of the other chapters including ***catch and store***, ***least change***, ***yield unlimited*** and ***renewable resources***. This chapter focuses on the pros and cons of energy efficiency and how it relates to other principles. As permaculture designers we don't have a principle that clearly asks us to be more efficient with energy even though it is a common phrase in the green energy[6] and green economy sectors. David Holmgren writes extensively about all aspects of efficiency[7] and has indicated that energy efficiency is primarily a matter of the two principles of ***obtaining a yield*** and ***producing no waste***. I felt the need to have a chapter exploring the subject of energy efficiency on its own.

We all love efficiency

You will hear energy efficiency talked about and promoted by physicists, engineers, economists, energy providers, governments, corporations and environmentalists. It appears to be a principle that could unite those of all political persuasions, but as we will see it is not as straightforward or as simple as just using less energy.

Energy efficiency at its simplest is the intention to lower inputs but have the same quantity of outputs, or to increase outputs with the same level of input. It could apply to energy in its many forms, such as materials, fuel, money and time, all as aspects of a system. If we insulate our homes we can maintain the same level of temperature comfort inside for less fuel, or we could replace the old fridge for a new more efficient version and keep the same cold temperature inside for less electrical input.

We can measure efficiency by calculating the energy input requirements and the work outputs. The most efficient is the system that does the greatest work output for the least input and minimal losses. For example, we can consider how much fuel is required by different types of transport to travel a set distance. The sums have been done[8] and we know that bicycles are the most efficient form of transport, even better than walking. Electric cars are more efficient than petrol cars and electric trains are more efficient than diesel trains. So next time you pop into town for that thing, get on your bike, and before you go make sure you have at least three reasons to go thus ***stacking*** in multiple ***functions*** at the same time.

Interestingly larger animals are more efficient than smaller ones.[9] This is because, as body volume increases, surface area increases more slowly and so a big animal loses less heat per gram of body weight compared to a smaller one. Thus, kilo for kilo, the larger animal can eat less. A tiny shrew might need to eat three times its own body weight every day. Wow, that would be equivalent to eating 200-300kg of food every day as a human! Despite our apparent efficiency with food we are the least efficient ape. This is partially because of our hungry brains and extra fat stores, but we have compensated this for lower efficiency with our social nature and collectivised activities such as hunting together.

In ecological systems we can explore efficiency as the flow of energy from one organism to the next,[10] for example, by measuring the energy gained by a plant through photosynthesis or the amount of energy gained by a herbivore from eating that plant or by a carnivore eating the herbivore. In every level of an ecological system, typically only 10% of the energy available is actually gained by the next organism, which sounds quite inefficient but it obviously works and is good enough. The other 90% of the energy is not absorbed or used in processes of respiration, heat loss, incomplete digestion or movement, such as a carnivore running to catch a herbivore. Efficiency for survival of a species is essential. If you spend more energy obtaining food in comparison to how much you actually gain from consuming that food then you are going to struggle to survive; in fact more than likely you wouldn't have achieved one of the essential criteria, a certain level of efficiency, for evolving in the first place. This overall decline in energy available as we go up the food chain is a good reason for shifting our diets towards being more plant based, because it means we can produce more food on less land as compared to eating animals, which need to eat plants first.

> *All living organisms are 'open systems': that is to say, they maintain their complex forms and functions through continuous exchanges of energies and materials with their environment. Instead of 'running down' like a mechanical clock that dissipates its energy through friction, the living organism is constantly 'building up' more complex substances from the substance it feeds on, more complex forms of energies from the energies it absorbs.*
>
> **Arthur Koestler, 1967**[11]

There's a better way to do it – find it

Energy efficiency has come a long way since it burst onto the scene in the aftermath of the early 1970s oil crisis. The drive to find ways to create better appliances has been fruitful. Industry and governments have loved it as it saves money, increases profits and is a vote winner for economic efficiency too. We now have fridges, freezers, light bulbs, cars, trains and planes all more efficient than before. A typical household fridge in 2020 uses 62% less electricity as compared to those in 1992, at the same time as doubling in volume.[12] It has taken scientists, such as Brenda Boardman,[13] to do the research and highlight the importance of government policy changes to push towards more efficient goods and buildings. One of those changes has been the introduction of energy efficiency ratings charts displayed on appliances. They are also available for homes but the take-up has been slow in the UK until recently due to a lack of enforcement. From 2023 landlords will be unable to let commercial properties that are below the E rating.

The strategy for these charts has been highly effective by raising awareness of the issues amongst the public and the gradual outlawing of the manufacture of the least efficient. This has been done in consultation with industry so that they have time, typically three years, to design, manufacture and market the next more efficient product. Changes such as these, alongside increased double glazing and the switch in home lighting towards LED lights, have reduced energy consumption. Electricity consumption in the UK peaked in 2005[14] and has been declining ever since and the efficiency drive is one of the factors that has contributed to that. It sounds great but there are key flaws in this drive towards energy efficiency.

Progress demands efficiency

The modern capitalist system loves efficiency because it drives progress, increases consumption and makes money. Every time a new energy efficient product is available then it can be marketed to us to encourage us to throw away the old item and get the new shiny replacement. Replacing goods too soon has driven up electronic waste, creating the birth of whole new recycling industries.

The energy consumption of the appliance is not the only energy involved. We also need to consider the energy requirements of its end of life and also the manufacturing of the product in the first place. This is called embodied energy, and is useful to help us make decisions about which product to use in a design as we should use the component with the lowest embodied energy. For example, straw bales, cellulose insulation and wool insulation all have a far lower embodied energy as compared to polystyrene or polyurethane-based insulation products. It has been estimated that 30% of all of the energy consumed by a building over its whole lifetime will be the energy used to make all of the parts and to build it in the first place.[15]

The drop in UK electricity consumption may well reverse as all of the electricity saved could suddenly be used to power your new electric car, your electric heating or your electric cooker as we switch away from petrol, diesel and gas. Furthermore electric vehicles have a far higher embodied energy compared to petrol vehicles, mainly due to the batteries and electronic components.

We can see efficiency in the development of fossil fuel powered machines: the engine, the motor car, the chainsaw and the tractor. They mean we can get somewhere quicker, cut down more trees per hour and plough more acres each day. To achieve more work in a day and earn more money in a day feels great, but at what cost? The earth and our communities are the cost of course. The straight rows of large-scale fields with hedgerows removed are more efficient than a random planting of a field with multiple crops. The big fast motorway is more efficient than the small winding road as we can get more people and goods across the country quicker. The bigger the scale the more efficiency there can be. This increased efficiency, as we are seeing, has pros and cons. At a household level it is more efficient, in regards to energy use for cooking, to buy tins of cooked beans rather than cooking them from scratch yourself.[16] At the same time the scale required to gain that efficiency for many products results in negative impacts on the earth and communities. If we are to apply energy efficiency then it must be combined with the permaculture ethics of ***earth care*** and ***people care*** and with other principles such as ***small and slow***. As permaculture teacher Starhawk has described amongst her principles ***energy is abundant but not unlimited***.[17]

If an appliance is hard to recycle or a house is still dependent on the use of fossil fuels, that means it is not sustainable in the long run and the item either needs further design or should be removed from our society. Even those LED light bulbs are tricky to recycle with their small fiddly parts. We need to apply the principles of ***produce no waste*** and ***use renewable resources*** in our redesign efforts as well.

The efficiency delusion

There is another side to efficiency, a darker tale to tell, and that was first worked out by William Stanley Jevons in a report he produced in 1865. As he described: 'It is a confusion of ideas to suppose that the economical use of fuel is equivalent to diminished consumption. The very contrary is the truth.' His hypothesis is known as Jevons Paradox, or more recently referred to as the Rebound Effect.[18]

Jevons was asked to work out how to make a steam engine more efficient so that less coal could be used to run the engine, with the intention that less coal would be required by industry. By making the engine more efficient it would of course be able to generate more power output per unit of energy input. But there's a catch. By making the engine more efficient there will be economic and behavioural impacts, such as the increased use of the engine or an increase in the number of those engines because it could now be run at a lower unit price. This drop in costs results in cheaper products and services and that can drive demand up. This would make the overall use of those engines increase and so the overall use of coal will increase.

We could see the same effect in household electricity use up until about 2005. Despite the previous three decades' shift towards energy efficiency measures and more efficient household items, the amount of overall electricity consumed continued to rise. This is thought to partially be caused by the increase in the number of household electrical items, their increased usage and increased size. As already described, electricity consumption started to decline in 2005 but only slightly.

Jevons Paradox is a worst-case scenario of the rebound effect, which is complicated and the extent of the theory is still debated. Only when there are ***limits to consumption***, an aspect of the ***fair shares*** ethic, will the overall energy consumption actually drop dramatically. Putting those limits to consumption in place has not happened as we are driven by consumerism.[19] The increase in electrical goods has added comfort and capacity to our lives and consequently we have more electrical goods than ever before. Even though LED lights are up to six times more efficient we have simply added more lights into our homes to make them even brighter. TV screens have become more efficient but now they are massive and can even occupy the side of buildings for advertising.[20] Energy efficiency is just not enough on its own. We need a whole systems thinking approach to reduce our overall energy consumption.

Turn it down, turn it off and keep it off

It is as if the gains made by energy efficient products are cancelled out by the increased quantity of products. Strategies to reduce the overall energy consumption are typically referred to as *energy conservation* or *energy sufficiency*.[21] These approaches argue that we need to use less, have less and use the lowest energy option, which is often to turn something down, stop using it or use an alternative. Instead of buying the new big energy efficient fridge, you could alter the fridge thermostat to slightly raise the internal temperature or even reduce your need for a fridge and buy a smaller one, or even question the notion of needing a fridge at all. I have lived with a cool cellar rather than a fridge for the last twenty years. There are changes in behaviour required, such as not buying too much at one time and using up what needs eating first. When I visit someone else's house I am often surprised by what they store in the fridge that does not need to be in there and how much they have in the fridge or sometimes how little.

A person focused on the energy efficiency of an item, such as a tumble dryer, might look for an energy rating of A+++, much more energy efficient than other tumble dryers on the market, or one already owned. A person focused on energy conservation might question the need for a tumble dryer at all. Identifying that the function of drying clothes might alternately be served by a clothes line with no need for the imported energy to run, nor the embodied energy to produce, a tumble dryer. The difficulty here is that changes on the scale required to lower energy consumption in terms of climate change become dependent on voluntary changes of behaviour by individuals, and that's not going to be fast enough.

Permaculture designers have achieved a lot in overall energy reduction through the ***relative location*** principle, the considered placement of elements so that energy use is minimised. We place the higher energy demanding systems closer so that travel is minimised, and consider how the flows of energy across a site can be utilised. We frequently challenge the need for modern services in the first place and often opt for very low energy input systems whilst still considering the ethic of ***people care***, such as the need for comfort. For example, building a Passivhaus, straw bale house or other low energy house will actually increase your experience of comfort in a home and yet your energy bills will be staggeringly low. There is evidence to show that those that live in inefficient homes use less energy, mainly due to the severe costs and futility of heating such a poorly insulated property. This may be a way forward in terms of energy reduction but not in terms of ***people care***. A simpler example of careful placement is the use of an insulated flask left next to the kettle. Every time I fill a kettle I try to put the right amount of water in but it's not always correct and there can be far too much at times. That excess hot water can be poured into the flask and kept hot for the next brew or reheated at much less energy use.

Another design discipline that has a similar approach to permaculture is that of Factor Ten Engineering,[22] or radical efficiency through integrative design, developed by Amory Lovins. It has its own set of principles. Lovins describes the concept as 'optimising buildings, vehicles, factories, and equipment as whole systems for multiple benefits, not piles of isolated parts for single benefits'. The process, he said, saves both energy and capital cost, and is clearly influenced by whole systems thinking.

Energy efficiency measures often assume that we need to keep the same levels of modern day living such as mobility and convenience, for example maintaining that we need to keep flying the same number of miles every year but use less fuel to do that. Energy conservationists say let's just fly a lot less as well and design our lives so that the need for extensive mobility is reduced. Our goal needs to be

One of Matt Ralston's highly efficient rocket stoves © Wilf Richards

overall energy reduction through both energy efficiency and energy conservation policies whilst also including the permaculture ethics. It is key that we get a balance between ***people care***, for example the affordability of energy and the energy needs of people, and ***earth care***, recognising that we must live within our one planet boundaries.

Can you handle it?

There is another energy efficiency principle focused on saving us time from the world of bureaucracy called ***only handle it once*** or OHIO for short. It was introduced to me by an engineering friend in the late 1990s in relation to the management of materials in our landscaping projects. If we have a large delivery of sand, clay, top soil or compost, we want to make sure the delivery is dumped in such a way that we only handle it once to get it to its final destination. It can be very frustrating when you realise the delivery has been dropped off in such a way that means you will need to move it sideways first. The worst case I have seen was a delivery of wood chips right in front of our access track at the end of the day preventing us from leaving the site. The delivery driver had not seen us and was presumably in a hurry. We were not ready to move it to its final destination and it was time to go home. Fortunately there were plenty of us to move it sideways and it was only a ton of material! I decided to add this ***OHIO*** principle into my permaculture teaching from about 2010. It was the first time I thought about adding and tweaking principles, which in turn led to me questioning the existing principles.

The principle of OHIO, promoted by Robert Pozen in his book *Extreme Productivity: Boost Your Results, Reduce Your Hours* (2012),[23] has been applied in many contexts including processing paperwork and emails, helping those with ADHD,[24] decluttering, time management, library cataloguing and assembly lines.[25] OHIO can become utopian idealism putting pressure on us to get things right the first time. We all have more work than time allows and so being perfect is an impossible aim, and do we even want complete efficiency all the time anyway?

The art of inefficiency

We recently experienced a classic mess up of OHIO when we were given a log cabin as part of an asset transfer. We were excited about the possibility of putting it up but knew it could be at least a year before we got round to the task. We discussed where it definitely would not go and that gave us what we thought would be a great place to store it. A place it would not be constructed but probably nearby. Twelve months go by and an opportunity arises to get the log cabin erected but in that time opinions and situations had changed. The perfect place to erect it now was exactly where it was being stored. It had to be moved again before it could be built! Maybe we can be content that we only handled it twice rather than three times.

Inefficiencies can give time for rest, contemplation and creativity. In the case of the log cabin the process of moving it again gave the chance to assess all the pieces, check it was all there and restack it in a way that would be more efficient for its construction, such as putting related components in the same pile. The processes of art, comedy, music and any creative practice can thrive in a place of inefficiency, a place of playfulness with no particular goal except fun and flow. These intuitive methods are hardly a fit place for efficiency. After all, rivers do not flow directly in a straight line to the sea. Given the chance I love a bit of idleness, a chance to sit and stare and see what emerges. A chance to wonder and wander, to see what is around the corner and not travel directly to our destination but instead dwell in the journey and the process. We need adventure and exploration too.

When we are designing, we must identify and understand our goals. Are we designing for minimum energy expenditure or maximum resilience or something else in between? We may find that we desire qualities and functions that are energy inefficient or require large energy investments. For example, the need for robustness to the difficult conditions of climate change could require a taller, wider bridge that can cope with larger quantities of flood water. Or maybe it is a reserve of energy that we need to ***catch and store*** to mitigate a disruption such as potential future economic collapse or electricity system failures. Having that increased resilience in a system could mean more expenditure of energy in the first place. The principle of ***every element serves multiple functions*** encourages material efficiency, but its counterpoint principle ***every important function is served by multiple elements*** encourages redundancy not efficiency.

We don't always want to be efficient. The most efficient food we have ever developed is refined sugar but you wouldn't want that to be your entire diet. Instead our bodies need the inefficient nutrients and calories from high fibre, wholemeal and raw as much as possible to maintain gut health.

So as we can see there are times when we want energy efficiency and times when we don't. I propose that it is adopted as a permaculture principle, and like many other principles, needs to be considered with care and context, and where relevant combined with the ethics and other principles when we are designing.

Everything Gardens

Written in collaboration with Carla Moss, Jo Holleran and Rachel Phillips

We all garden

Bill Mollison coined the term ***everything gardens*** in his book *A Designers Manual.*[1] He referred to everything having an effect on its environment and how everything makes its own garden. He asked us to observe how our local plants and animals change their ecosystem and to find the allies that could help us. He also talked a lot about rabbits. There's a picture to go with his first exploration of this principle showing a kangaroo 'pruning' a tree, a hen 'digging' and rabbits 'mowing' the grass. The principle is still included on the Permaculture Association[2] website too.

The key point is that it is not only humans that garden. We think of ourselves as the only gardeners: sowing our seeds, planting out, weeding, harvesting and digging. This principle helps us to see the role and effect that all other living beings have on their world. We are not the only gardeners. We are not the only ones affecting our surroundings. We need to recognise and acknowledge the massive impact that we have on our environment. We modify it enormously, but imagine if your choices made no impact on anything. That the things you did and said and thought and felt had no repercussions. Not a single one. That is just not the case, no matter how lightly you wish to live on the earth. Even when a plant or animal is dead it decays and contributes to the life of bacteria, fungi, worms. Literally everything is gardening all of the time.

Everything is alive

Do we literally mean everything? The early applications of this principle focused on the effects that animals have on their local environment but this principle encompasses so much more than animals. Plants, fungi, microbes and even mountains, the seas, the moon and the sun also 'garden' and modify their environments. We can view everything as alive, constantly changing its immediate environment and the surrounding species even if imperceptibly slowly.

Gaia Theory,[3] which had a big influence on the development of permaculture, states that organisms co-evolve with their environment by influencing the chemical and physical parts of their surroundings which in turn influence the biology of their environment. For example, as organisms breathe and respire they change the chemical makeup and temperature of their environment. The emergence of photosynthetic bacteria millions of years ago resulted in an oxygen enriched world. The eruption of a volcano spews vast quantities of sulphur dioxide into the atmosphere cooling the surface of the planet.

Taking it to the next step we can consider everything as literally being alive, rather than just influencing its surroundings. This belief, called Animism,[4] is common amongst Indigenous peoples but not all permaculture practitioners will believe in it. That belief might include the concept that everything has a soul, hence for many Indigenous peoples there are many spirits and deities, each connected to and embodied in a species, landscape and ecology around them. Animism gives those entities agency and a power, which in turn requests respect for everything.

Even words can be seen as alive as they come into existence, change over time, affect the culture and then may one day disappear from our language, or even be resurrected. We may see objects made by people as alive as they come into existence, provide a need or a want, prompt emotions such as desire, create connections in the mind through memory and inspiration; they might cast a shadow, stop a breeze or create a microclimate, and maybe decompose back into the earth one day.

Animists do not see themselves or any other living being as separate from the environment. Everything is connected and ***integrated.*** Remnants of this belief can still be seen as place names on maps, such as Holywell, which is a common name given to a source of water that would have been regarded as sacred and life-giving. This principle invites us to deeply connect with our surrounding natural world and understand the power that everything has. Whether you believe it to be 'alive' or not, ask that element in your landscape what it 'thinks', what it does, what it knows, what it needs and how it wants to be in this world.[5]

Everything modifies its environment, for better or worse

Birds gather sticks for nests and scratch away leaves looking for worms; rabbits nibble the grass, squirrels harvest hazelnuts, jays bury acorns, fungal mycorrhizae trade nutrients with tree roots, plant roots crack open rocks, rain dissolves soluble rock and moves soil, mountains slowly grow and diminish. Everything has the power to modify its environment in some way.

From an evolutionary perspective the purpose of modifying your environment is to make it more suitable for yourself. If you get this right then you will create a better habitat for your species and then you are more likely to survive. The modifications that you make to your environment though might not just benefit yourself. They may also result in benefits to other neighbouring species too who will take advantage of the changes that occur. In some cases the modifications made by a species will be detrimental to itself in the long run.

Some plants, such as the beech tree, have allelopathic properties, which means they release chemicals into the surrounding soil that suppress other plants from growing. This could be through the roots or the leaves that drop in the autumn. Either way the plant is actively modifying its environment, almost in a selfish manner, to keep all of the available resources in that vicinity to itself.

I have seen the assumption made about this principle that it is always about a species modifying its environment for its own benefit. I don't think that is always

Under the beech trees not much grows © Beth Silverbirch

the case. Yeast will turn sugars to alcohol resulting in the ultimate death of the yeast, some will survive in a sleep like state to be woken up again and breed when fresh sugars arrive. This ***pattern*** applies to most bacteria and microscopic fungi. They will burn through the intense availability of particular nutrients to thrive and multiply until that resource is diminished. A pattern unfortunately common to some human behaviour too.

Some species will benefit future species. This is common in patterns of ecological ***succession***. The spiky plants such as thistles, brambles, hawthorns, gorse and blackthorns play a vital role in keeping herbivores away, which in turn means that tree species gain an advantage and become established. Once those trees reach a certain height, they will cast shade so strong that the spiky plants can no longer survive.

We can see a similar yet far more complex situation in large herbivore groups when allowed to roam their full territory. They may debark trees, even push them over in the case of elephants to access the leaves and younger stems. This might result in the death of the tree but might also increase the likelihood of grasslands to thrive in the spaces now open without a tree canopy. Some of those herbivores may move onto new territory to continue their browsing habit and in turn encourage the grazers of grass to follow them. With time the ***succession*** continues and the trees will grow once more.

Deciduous trees will cast shade on the ground around them in summer which will keep the ground cool and minimise evaporation of water from the soil, a direct benefit to the tree itself. Some plants can fix their own nitrogen through a

Wildflower nigella thriving with wood strawberries and mint © Jo Holleran

beneficial relationship with specialist bacteria in or near their roots. This nitrogen will support the growth of that plant but when the plant dies it will also add additional nitrogen into the soil making it available to the next plant.

As humans we have been modifying our environment, although very much in the wrong way, so that we make it worse for ourselves and many other species. Many of our modern day actions pollute and destroy the surrounding habitats. Our cars fill the air with pollutants that make it harder for us to breathe, our sewage systems ruin the water that we need to drink, and our poisonous chemicals destroy the soil we need to grow food in.

Work with nature, become a gardener

We need to become gardeners ourselves. Let's get to know the plants and animals in our environments and ***observe*** what they have to teach us about gardening the landscape. Which ones can help us grow our food, fuel and fibres in a more natural way? How can we support the habitats that these experienced non-human gardeners need to thrive?

Maybe it's ensuring you have a healthy hedgerow to encourage the birds into your market garden. Maybe it is the robin from the hedgerows that pops down to the bare earth and eats an insect larva that would have damaged a crop. Maybe it's the worm that turns the soil and keeps it aerated. Maybe it's the cat that catches the mice. Or the thistles left at the edges that attract the aphids away from your tender crops. All of these actions also fit within the ***least change*** principle too.

Companion planting[6] is about utilising the behaviour of one plant to benefit another, typically the addition of a strongly scented wildflower to benefit a plant crop. For example, the addition of calendula that will attract aphids to their sticky leaves and flowers, keeping them away from neighbouring crops. They may also attract ladybirds and hoverflies, which prey on the aphids. The amount of good evidence for companion planting is unfortunately very small. What is more likely to be effective is simply the ***diversity*** and minimal monoculture, thus keeping pests to a minimum as their food supply is limited. Another approach is to allow the wildflower companions the opportunity to self-seed or spread. They will reappear if the conditions are right and thrive if the conditions continue to be right for them. Allow the plants to make their own choices and ***observe*** how they are helping you. Some plants enjoy being more on their own whilst others clearly seem to tolerate or thrive in the company of others. And if any appear a bit too dominant then maybe they just need harvesting more.

Most gardeners seem to hate slugs. I implore you to research slugs.[7] There are hundreds of species. Some slugs are carnivores and will eat other slugs. Many are important for the processing of dead plant and animal matter, a key role in returning vital nutrients to the soil. If you find they have been eating your crops then identify which crop and maybe stop growing that crop for a while, as you have been feeding and breeding that variety of slug. If you are lucky enough to have snakes in your patch then look after them as they love slugs.[8] Although do they like the same slugs that are eating your crops?

We may choose to not dig our gardens or at least minimise disturbing the soil, to encourage the delicate mix of bacteria, fungi and insects in the soil to carry out their roles in supporting plants to thrive there, another example of ***least change*** as well. We might add compost or mulch to the soil to feed those critters, or even deliberately add particular fungi that have been shown to support particular plants. We may also choose to involve other people in our garden, volunteers that want to connect with other people, that want exercise and some experience of working the land without having the responsibility to manage it all.

Persecuted creatures

The mole[9] seems a particularly persecuted creature in many a rural UK field. They are considered pests by many gardeners and farmers alike. The fox, their main predator, is persecuted too meaning the moles maintain high populations unless a mole catcher is brought in. Typically in a rural pasture you will find rows of dead moles hung on the boundary wire fences. In the recent past we turned them into fur coats, an unpleasant example of ***the problem is the solution***.

The main crime of moles is creating molehills, which in turn can reduce the yield of hay or contaminate the hay with soil particles, which may spoil silage or make the hay less palatable to livestock. As for those that deeply care about having an immaculate lawn, then the mole is the number one enemy.

On the positive side moles can be introduced as a biological control of cockchafer beetle larvae, which have a particularly high impact on grass quality. Moles also help to keep the soil healthy through the creation of tunnels that can go down five feet and help with drainage, especially on heavy clay soils. We can also see molehills as nurseries for wildflowers. Their main diet is earthworms so when we see their molehills in a field, we also know that the population of earthworms

Mole gardening, even in the snow © Wilf Richards

must be healthy there too. They also feed on slugs. If we switch our perspectives about some species then we could see them as a solution rather than a problem, as we understand the positive ways in which they garden.

The rights of nature

In her mixed fruit orchard, Jo has seen roe deer with their fawns contentedly eating the fallen fruit. Windfall apples are graded every few days, those suitable for human consumption are gathered up and used for cooking, preserving, or cider making. Unsuitable apples are left on the ground for a further few days, to give non-human gardeners the opportunity to tidy them up. They will provide nutrition to a diverse range of insects, birds, and mammals such as the deer.

Other non-human gardeners regularly contributing to the orchard include countless types of bug, ants, worms, slugs, bees, wasps, hornets, blackbirds, starlings, robins, redstarts, wrens, woodpeckers, red squirrels (especially in the hazelnut and walnut department), stone martens (the plum department), badgers and foxes (the cherry and worm department), moles, field mice, edible and garden dormice, three cats (the small-furry population control department) plus opportunistic families of wild boar.

All species should have the right to do their gardening in their own natural way, being themselves, and where we can find a way to work alongside our fellow gardeners we may also find ourselves saving time, effort, money and energy, and in so doing we also enact ***fair shares***.

Functions and Elements

Written in collaboration with Chris Warburton-Brown, Arnaud Fache and Rakesh Rootsman Rak

The two cousins

There are two complementary and closely related principles about functions and elements: ***each important function should be supported by many elements*** and ***every element should provide many functions.*** These were presented as two of the key principles in the book *Introduction to Permaculture* (1991),[1] but they were first expressed in *Permaculture One*[2] back in 1978 as 'each element serves several functions in the ecosystem, and each function is common to many elements'. Clearly right from the start of permaculture, increasing functions, elements and connections was key. John Quinney had written about them in his 1984 article[3] helping to bring them to the surface, referring to them as ***multiple functions for single elements*** and ***multiple elements for single functions.*** Quinney shares many examples including the New Alchemy Institute's water storage system within their greenhouse that provided multiple functions for fish ponds, warm irrigation water and nutrients for a hydroponic system. Also he explains that by using multiple elements for any given function provides an insurance policy for failures in a system.

Various other manifestations of these principles have popped up over the years, such as from Patrick Whitefield's book *Permaculture in a Nutshell* (1993):[4] ***every need should be met from many sources*** and ***every plant, animal and structure should have many functions.*** And then there is Sarah Spencer's tweak[5] which focuses on needs: ***each important need should be supported by several different strategies*** and ***each activity or element should fulfil several needs.*** These two principles, in their various forms, have been a consistent backbone of permaculture from the early days. They are easy to understand and really help to shift our perspectives and actions towards ***resilient*** and stable systems. David Holmgren incorporated the two principles into his ***integration*** principle in his book[6] which is very appropriate. This also means that some permaculture designers, focused solely on Holmgren's work, might not be aware of these two original principles.

The dictionary definition of a function refers to an activity, purpose, task or work. There are a lot of definitions of the word element but they generally refer to a component, entity, substance, constituent, ingredient, forces or even niche.

The choice of these words, elements and functions, can seem too technical and mechanical to some. They conjure up images of mathematical formulas, computer language, chemistry labs, weather forecasts and magical forces. They hold the risk of detaching us even further from natural systems that we are supposed to be already part of and integrated with, but the words have stuck as part of the permaculture jargon for many.

As permaculture designers we see the elements as the components of the systems we create and the functions are the relationships between those components. Or we could see the elements as the 'whats' and the functions as the 'whys' and 'hows'. These two principles are about connection and a recognition that the systems we create are complex just like a natural system should be. Functions could be goals, purposes, needs, criteria, activities or reasons. It is really important to remember that we are always one of those elements in our designs and we are never separate from the landscapes we dwell in.

The line between element and function can blur. Water might be an element but watering could be a function as functions can also be viewed as needs in a system. For example, our food crops (element) need watering (function) and we may use water from the rainwater tank (element). The blur between elements and functions can be a frame of reference issue: we need to take into account the specific context and our perspective of the situation. The context could influence the importance of one function over another and this could change in a different circumstance or at a different moment in time.

The watering of our crops might come directly from the rain but also possibly ground water, swales, water stored in the soil, a river, a rainwater harvesting system or the mains tap. We may use all or many of these water sources. This is an example of an ***important function being supported by many elements.*** For further examples and understanding of functions I recommend Angelo Eliades' Deep Green Permaculture website.[7] From the other perspective, the plant we are watering could be an element providing many functions, which could include food for people, food for livestock, a habitat for birds, a windbreak, a soil fertility builder and medicine.

Permaculture teacher Rakesh Rootsman Rak,[8] with his love of music and electronic skills, has created a mobile bicycle-powered sound system, which of course has many functions and many elements to it. The elements are the many components carefully put together to get the whole thing to work. The functions might include creating electricity, playing music, getting exercise, getting people dancing, demonstration and education about how the sound system works and as a starting point for a conversation about permaculture or any other topic of interest of course.

Rakesh and his bicycle-powered sound system © Wilf Richards

Resilience and redundancy

The two principles about functions and elements play a key role in the ***resilience*** of any system. If a key function can no longer be supported because it was entirely dependent on a single element, and that element has failed, then the system may collapse. We can consider having a plan B in place in advance, or have multiple elements as backups to that key function. One of these strategies may be to create redundancy in the system. In fact some permaculture teachers refer to the principle of ***every function should be supported by many elements*** as redundancy or as functional redundancy.

Resilience and redundancy are related but not the same thing. ***Resilience*** focuses on the ability for a system to bounce back or keep going despite a shock or fault. Redundancy is putting something in place, such as duplicate elements, to prevent that failure happening in the first place.[9] Definitions of these two words vary according to the context and professions that are using them. Here we will focus on redundancy as there is a whole other chapter on resilience.

Rakesh Rootsman Rak has a background in IT systems where he first learnt about redundancy. Part of testing a new IT system is to try and break it in advance to see where the weaknesses might be. Once identified, back up systems, also called redundant systems, can become one of the solutions, so that when the failure occurs in the main system the redundant parts come into play. These might be spare components that are rarely needed but provide an essential resilience to a system. Rakesh refers to this as the principle of 'eliminating single points of failure'. Depending on the context, this might be done by building in additional connections, a ***diversity*** of alternate components or alternative pathways. Another principle Rakesh has brought in from working with electrical systems is 'designing for surge and pulse'. In electrical terms this is looking out for possible extremes of too much power or too much voltage in the system and needing to install contingency systems or controls on loads to prevent possible failures. Rakesh has carried these principles into his work with forest gardens and permaculture. This might be in terms of anticipating potential seasonal extremes such as flood or fire, and knowing what is sacrificial in the system.

System redundancy could also include emergency electricity systems such as those found in hospitals, where you really don't want the mains grid electricity to fail and when a power cut does happen, the in-house backup system kicks in and lives are not lost.[10] We can also include the spare car tyre in the boot or packing the spare inner tube on a bike ride as examples of redundancy. The need to include redundancy will increase with the level of importance of the function that is being considered. Functions based on our survival such as supplies of food and water need to be given higher priority when it comes to adding in redundancy. Examples include multiple sources of water and a wide variety of crops to minimise the likelihood of loss to pests and diseases. For electricity systems we have, as a society, a range of supply types including solar, wind, hydro, gas, coal and nuclear, providing us with a steady supply. We also need a diversity of income streams, colleagues that know how to do our essential tasks when we are ill and as permaculture teacher Patricia Allison put it in one of her handouts,[11] more than one friend.

Putting in backup systems is not always a good idea though. Apparently Galileo first noticed this in 1638 when a safety system had been put in place that actually made the system less safe. Contrary to expectations, a precautionary measure can cause a disaster. Galileo's example was when a third midway support was placed under a long horizontal stone column as a precaution.[12] The result was an end support sinking and the column breaking in the middle. Many more examples have been found since, including the explosion prevention device that caused an explosion and a filter installed in a nuclear reactor to increase safety that broke, nearly causing a meltdown. These are rare but worth being aware of. Adding in backup systems adds complexity to the system and the more complex the system the less likely we are going to predict the interactions and knock on effects.

Functional analysis

When we are considering the other key principle, that ***every element should provide many functions***, we need to do some research. An element in our system might be a plant. We need to know more about that plant to consider its best use or placement. We may wish to know the plant's life cycle (e.g. annual or perennial), its shape and size, its climate requirements, ideal light needs, water tolerances (damp, drought etc.), soil preference as well as its many possible uses or functions (food, fodder, shade, seed, soil stabilisation, medicine etc.). One method we can use to map that is called functional analysis, sometimes referred to as an input-output diagram. Typically we need to know the inputs or needs of the element, the outputs or products of the element and the intrinsic characteristics, which are internal or context specific. Bill Mollison first demonstrated this tool with a chicken.

This design tool can bring to our attention where an input is actually an external energy or extra work, and where an output is a wasted resource or pollutant. More is written about pollutants and wastes in the ***produce no waste*** chapter. Here though our focus is on the functions of the element we are considering and consequently where that element can be best placed. I particularly like the example on the Deep Green Permaculture[13] website where the writer, Angelo Eliades, describes a willow tree's form (shape and life cycle), its tolerances (e.g. soil preference) and its many many uses. This leads to the placement of the willow at the edge between a river and pasture where the tree can provide riverbank stabilisation, filtration of nutrient runoff from the field and wind protection, shade and fodder for the livestock. This application of the principle focuses on developing an ***efficiency*** in the use of the element and how elements relate to each other, thus using ***relative location*** too.

Complex systems

Seeing the world in terms of functions and elements shows the complexity of connections. In all systems there are many functions and many elements, all ***integrated rather than segregated***. For example, when we consider a woodland, there is a wealth of plants, fungi, animals and microbes all connected in the web of life, giving and receiving and exchanging energy. As ecologists we see those complexities and we avoid creating systems that are focused on a single element for a single function, such as a wheat field solely for the purpose of food for sale. Instead we shift our perspective towards polyculture: seeing the hedgerows, the birds that visit, the microbes in the soil, the animals that pass by, the wildflowers amongst the wheat and the communities in our neighbourhoods and so much more. We want to shift towards that ***diversity*** and so we add in trees, support the wildlife and add extra crops. We want to see the range of possibilities and purposes that any habitat offers. A wetland is not only a place for wildlife but it is also a carbon sink, a store of water, a flood protector and beautiful too.

Mapping functions and elements

In 2019 on one of my permaculture courses, we developed a way of mapping the relationships between functions and elements. Maybe someone had done this before but it seemed like a first to us. This was at the end of the course during the group design phase. The group, Mike, Jen, Pete and Katie, were designing a community garden for Sherburn Hill on the east side of Durham City.[14] The design had the potential for many elements. The community had also identified many functions that they wanted the space to fulfil. So the team wanted to visually show all of the elements and functions and explore how to prioritise

		Functions												
		Community development	Sustainability	Sharing	Education	Generate Produce/enterp	Friendship	Accessibility	Biodiversity	Wellbeing	Shade and shelter	Participation	Play	Creativity	Seating
Elements															
Raised beds															
Big Polytunnel															
Entrance Polytunnel															
Gazebo/Sheltered area/seating															
Chicken coop/storage area															
Office/shop															
Bait cabin															
Storage containers															
Arts and crafts															
Sensory garden															
Pond															
Lawn area															
Playground															
Orchard															
The mouse hill															
Wildlife area															
Shared Allotment															
Front/side garden															

Functions and elements matrix original ,2019 Durham PDC © Wilf Richards

which elements would best fulfil the community's many requests. It was at this point that I suggested a spreadsheet as an option to cross-correlate elements and functions but then one of the participants added colour coding. It was one of those moments where we felt we had collectively come up with something very useful. The functions and elements matrix emerged.

When I used this matrix tool in my design for rest[15] I added in a scoring layer to provide a further nuance beyond just three colours. In that design I was looking at a range of activities (elements) to fulfil the various criteria (functions) that I used to define rest. This enabled a further level of analysis. I have also used this method to map out progress on the development of this book. The chapters being the elements and the criteria for each chapter being the functions.

In 2020, Arnaud Fache[16] developed another variation of mapping functions and elements in one of his permaculture diploma designs when exploring the placement of furniture and space use in his flat. This version was inspired by the *web of connections* mapping method that features in Aranya's book *Permaculture Design: A Step-by-Step Guide* (2012).[17] Rather than only focusing on connections between elements Arnaud's version showed functions as well.

In the image (right) we can see the elements are on the left hand side e.g. bed, sofa and the functions are the activities on the right in blue. There are yellow lines that connect the various elements and functions together. The elements and functions have also been put into an order of importance, the top ones being the most important. This method provides the opportunity for multiple connections to be shown. Arnaud acknowledges that this tool could be improved even further by using a colour code for the lines to indicate the nature of the connection, or using different line widths to express the strength of the connections.

This is a vast improvement on the potential chaos of lines that can exist in a *web of connections* map. Whichever method you use then aim to get at least three elements for every function and three functions for every element. That way you will see the complexity of the system you are designing shift away from the monoculture and towards the polyculture.

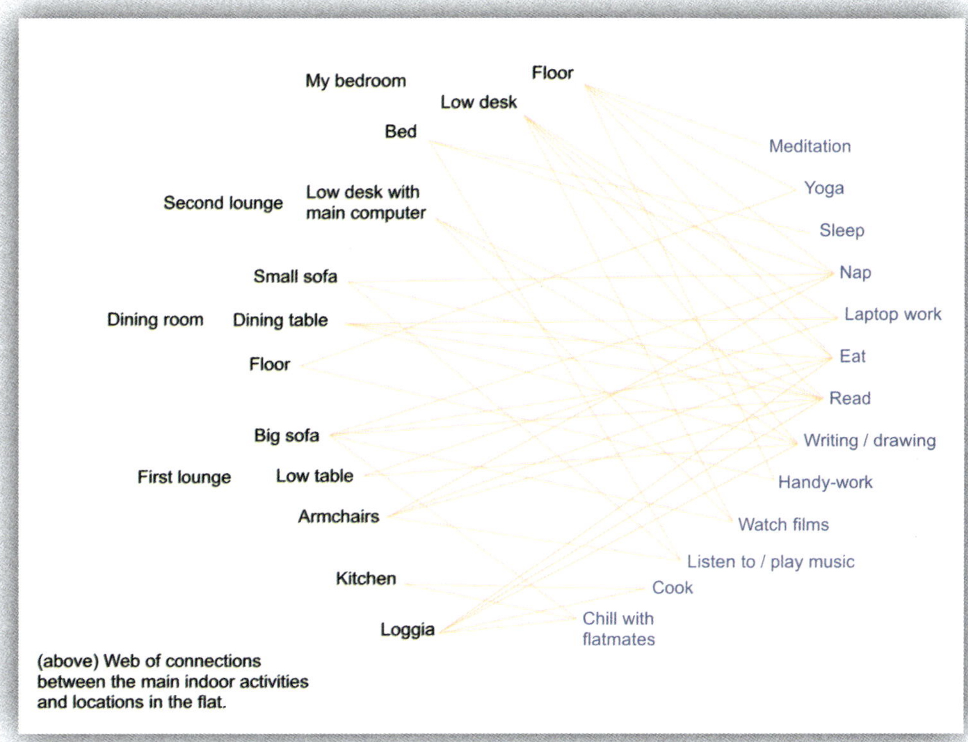

Arnaud's functions and elements web, 2020 © Arnaud Fache

Another method of mapping functions, elements and systems together is in a table, also from Aranya's book.[18] So we can see there are multiple ways to map out our functions and elements and the interactions between them.

Whichever method you choose, there is a nice overlap here with ***patterns to details***, in the sense that when we explore functions they could be pattern-level goals or detailed context-specific criteria, and at the same time our elements combine together to form pattern-level systems. We can find ourselves jumping between levels of perspective, from a pattern level to a detail level and back to pattern level.

Is stacking different?

You may have already noticed that there is another chapter focused on ***stacking***, a term often used to refer to multiple functions. The two can overlap but stacking has a wider definition and a different origin for me so required a separate chapter. Stacking refers to the observation of ecological ***succession*** and applications such as forest gardens. I think using the term stacking to refer to multiple functions is a form of permaculture slang and certainly an evolution of its original meaning. We might refer to *stacking* purposes, functions or goals. When it comes to teaching permaculture then this can get confusing for someone new to the concepts. Despite using this slang myself I am trying to use the term multiple functions when referring to purposes, needs, criteria, goals or tasks. When I refer to the placement of plants in a system, for example in a polyculture or forest garden, then I call that stacking. I think there is a case for distinguishing between these two terms. Your forest garden might contain a hundred plants all carefully placed and stacked to enable each plant to thrive in its best niche. The functions of the forest garden though might be food production, biodiversity, income, pleasure, mental wellbeing, carbon sink and work. You can read more about the differences and overlaps between these terms in the ***stacking and succession*** chapter.

Integration Optimised

Written in collaboration with Tobias Sonrisa Sturmer

Core permaculture

Right from the start of permaculture[1] we can read about integrated cropping systems, developing complex mixed perennial planting schemes of multiple ***stacked*** layers integrated with animals, and us humans integrated with the whole thing. This was a reflection of the understanding of natural systems where species do not operate in isolation but as part of an integrated ecology with interdependent relationships. Permaculture itself is an integration of many ways of being, many disciplines and many approaches, such as the integrated holistic approach of systems thinking.

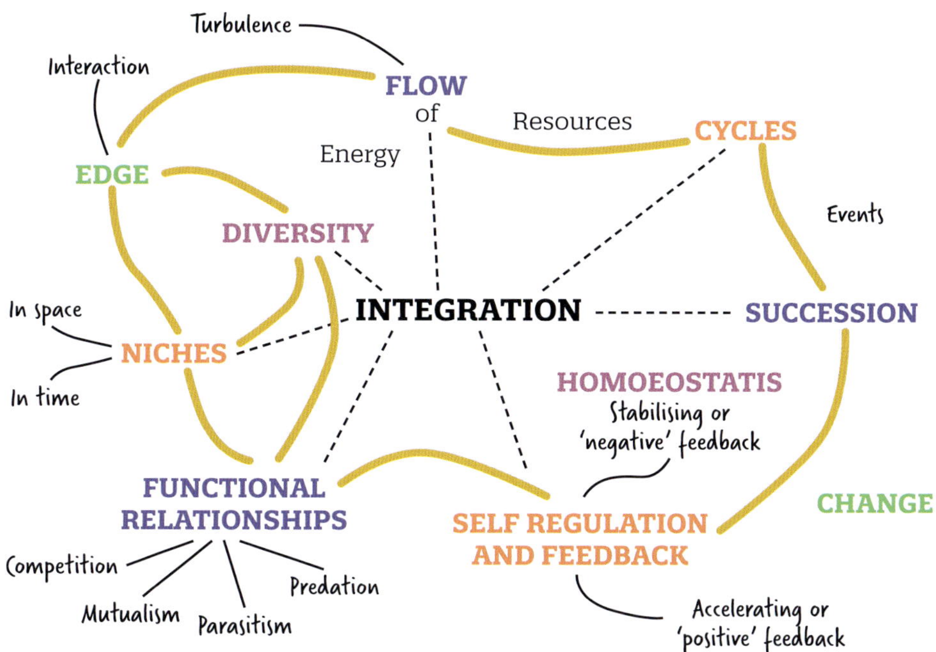

Tomas Remiarz's principles of ecology map, integration is central[2]

Several of the early principles are focused on integration, such as ***cooperation not competition***, ***edge effects*** that explore where habitats, resources and species integrate, and as discussed in the last chapter the two principles about functions and elements are all about connection and integration. In his 2002 book[3] David Holmgren clearly explains how the functions and elements principles led to the development of the principle he coined as ***integrate rather than segregate.*** I can see how other principles have connections to integration too, such as ***work with nature*** (integrating ourselves), ***relative location*** (careful placement to get the right level of integration), ***stacking*** (similar but typically considering vertical integrations), ***produce no waste*** (integrating wasted outputs to become new resource inputs), ***self regulation*** (integrating feedback into the system), ***patterns and details*** (integrating the fine points with the big picture) and ***diversity*** (increasing the amount of potential integrations). Integration is such a key goal of any permaculture design. Maybe it is the principle of principles!

The reason for this, as David Holmgren explains, is that permaculture designers have a focus on the relationships between things rather than just the things. The placement of elements influences the level of integration and when it is just right it adds to the ***resilience*** and ***efficiency*** of that system. Permaculture designers create systems with multiple ***elements*** and integrated ***functions.*** We like ***diversity*** and we like to find ways to neatly fit our elements together so they can mutually support one another. This can increase ***yields*** and reduce work when we get it right.

Holmgren shows how ***cooperation*** and our focus on ***functions between elements*** are key to enable steps towards integration. He describes how our society has focused on segregation for reasons of simplification and as a reflection of the superabundance of resources. I would also argue that it is for reasons of order and control, such as the saying *divide and rule.* Segregation can be a tool of the powerful to maintain their power.

One of the key anchor points in Looby Macnamara's Design Web is ***integration.*** It is at this stage in using this design process that we would typically bring together all of our thinking up to this point. This could be a reflection on all of the helps, ideas, patterns and limits that influence the vision and how the principles have influenced your thoughts. Now you are ready to put together an integrated action plan.

Integration in nature

Permaculture designers state that the connections between things are as important as the things in themselves. Taking this a bit further, we might realise that there really is no such thing as a thing in itself, this is a reductionist concept. Everything is subject to conditions and relationships. There is no isolation and separation in nature and this is a perspective we need to cultivate and remember. For too long we have seen human society as separate from natural systems and the earth. Let's start with our own bodies. Our mind is not separate from our body, it is an aspect of our body. We are not only human but actually contain more microbial cells than human ones. Most of these extra bugs that live within us and on us have benefits including protecting us from microbes that might harm us, supporting our immune system, converting food into digestible components and cleaning up our guts. Our microbiomes are totally integrated as part of our digestive and immune systems. When we regard only a part of our digestive system it can be helpful for diagnosing a particular health condition but it also stops us from seeing the complex relationships within our guts and connections to other parts of the body.

Complex forms of life were only able to emerge after one type of cell integrated into another type of cell. These were the energy converters such as mitochondria and chloroplasts, originally independent and free living. This is known as symbiogenesis,[4] the process of primitive prokaryotic cells joining forces to become complex eukaryotic cells. Only once this evolution had taken place, about 1.6 billion years ago, could the possibility of plants, fungi and animals begin to emerge.

We know from scientific studies and observations of nature that nutrients and chemicals move between rock, soil, air, water, microbes, fungi, plants, animals and all life. Key processes such as photosynthesis, respiration, digestion, transpiration and diffusion are all critical in supporting those movements. Any given species needs another species to enable its survival. Interactions, connection, exchange and integration are all around us. We live in a holistic world. The recent studies around connections between trees and fungal mycelial roots are extraordinary, showing exchanges of chemicals as food, communications and mutual support.[5]

> *It will not be possible in this integrated world for your heart to succeed if your lungs fail, or for your company to succeed if your workers fail, or for the rich in Los Angeles to succeed if the poor fail, or for Europe to succeed if Africa fails, or for the global economy to succeed if the global environment fails.*
>
> **Donella Meadows**[6]

Back to the land and finding community

A common dream many of us have is to connect with the soil, our food, natural landscapes and other people to fulfil an instinctive need for community and a harmonious way of living with the earth. It is a step towards being indigenous once more. We miss it because we need it for our survival and for the feeling of being whole. The development of community gardens, small farms and intentional communities are common amongst permaculture designers. Our projects grow what we can for food, fuel, medicine and fibre, utilising site resources such as rainwater, and involving people from the neighbourhood, sometimes through volunteering or offering a service such as mental wellbeing or nature connection.

It is all about integration. These are attempts to move away from personal isolation, supermarket dependency and the industrial society that segregates us from nature. We can bring an end to kids not knowing where vegetables come from and not knowing how to eat them, of isolated farmers in their tractors, long complex supply chains and massive waiting lists for allotments.

Direct access to the raw ingredients of food is a great stepping stone to our integration with this earth. That stepping stone might be an allotment, or volunteering at a community garden, or buying veg from a local grower. It might also be attending an event at your local community centre, joining a local club or actively encouraging the inclusion of the marginal within your activities. We know that through the processes of working together and playing together that we can encourage ***cooperation*** and build ***resilience.***

And yes, these are steps towards complexity. At Abundant Earth,[7] we have an integrated economic system so that all income streams, of which there are typically fifteen at any one time, go into the same cooperative pot enabling a steady and ***resilient*** income. Compare this to our original set up of multiple self-employed individuals each with their own boom and bust cycles. Our marketing is largely integrated too, which enables customers to see what else they could buy from us. Someone may start with us as a willow basket course participant but then migrate sideways as a veg box customer and then attend a permaculture course or a volunteer day. This ***diverse*** mix of land based and people based services is common on permaculture projects. Our smallholding is also ***diverse*** with many integrated land types, including forest garden, market garden, veg box packing area, pasture with fruit trees, hedgerows and woodland. There are many interactions between these different zones. Maybe you have experience of a similar set up that can also provide security, mutual trust and ***resilience*** such as a co-housing project, local cooperative business, interdisciplinary academic activity or a community project.

Integration of the various parts within our individual selves is also key to integrating externally. This might include coming to terms with a trauma in our lives, understanding the integration of our mind and body or mending external broken relationships through internal forgiveness and compassion. Knowing that some parts of ourselves are a bit broken, yet accepting and loving them all the same can be an important part of our journey through life to make us feel whole again. Processes might include meditation, mindfulness and therapy as ways to *garden the mind* and keep the ego in check. According to Zen Master Thich Nhat Hanh, the wisdom of non-discrimination is the letting go of any feeling of superiority, inferiority and even equality, as all three come from a place of ego and separation as self. In its place is the wisdom of inter-being. He says that we physically embody this wisdom: having two separate hands, which do not compete, that do not look down on or up to each other, and are not concerned about feeling 'equal'. And then there is also the wisdom of discrimination called *viveka,*[8] which is defined as the ability to differentiate and thus orientate ourselves.

Segregation bad, integration good?

Segregation at its worst in our cultures includes war, apartheid, racism, class systems and ghettos. It is enforced separation, power divisions, power over others and behaviours leading to prejudice and inequality. All of these types of segregation in our human culture are clearly disturbing, undesirable and horrible.

Then there are shades of segregation, pros and cons with complexities of clashing needs. These shades of segregation may be cultural protection, such as faith schools or countries with strong borders trying to manage what enters and leaves, like an immune system on constant alert. It may be prisons keeping people who have harmed others separate to minimise the possibility of them doing harm to others again. It might be sanctions placed upon a country to economically isolate it and to let that country know that its behaviour is not accepted by the majority. It could be women fighting for their right to female-only spaces. In our communities there can be examples where segregation starts to look more positive, where it is convenient and ***efficient***, such as schools looking after kids whilst the parents are out to work, an old people's home providing specialist care, or a college providing focused academic disciplines. Dividing our lives up into zones and sectors makes management far easier but at the risk of losing connection between those separate parts.

Our educational systems are full of segregation: a curriculum of neatly separated subjects; a school with separated ages, sometimes with pupils of the same gender and often wearing the same clothes; a considerable disregard of the body and its preference to move rather than sit, an obsession with seriousness and regurgitating as opposed to playful, fun and creative exploration; a grading system with a disregard for the many different learning types, and an overemphasis that we are preparing a young person to become a worker rather than a human. Which child is going to leave our schools after more than a decade of formative years spent under such a segregating regime and not be competitive, insecure, stressed and cognitively dissonant?

There are some positive segregation situations, such as keeping livestock away from our veg, the isolation of someone who is ill with a highly contagious disease, and the need for time on our own to recover from excessive socialising. Another example of positive use of segregation balanced with integration is the division of roles and responsibilities based on individuals' approach and motivation within a team. It can be very empowering if a clear role is agreed by the group and someone takes on that role knowing they have some power to make relevant decisions within their remit. It doesn't mean that person is working in isolation. Their role will still need to be defined in terms of neighbouring roles but there can be a freedom of decision making within an agreed boundary. There are various forms of governance that encourage such care, including sociocracy and ***cooperative*** structures. These types of approaches can minimise unnecessary debate and arguments, and allow individuals to thrive in partial autonomy within a supportive larger collective of people. They require courage to segregate to the right level, knowing that some degrees of separation allow room for growth and freedom with responsibility.

I question the notion that we live in a society that only focuses on segregation. I would argue that capitalist society loves certain types of integration. I see a global society awash with communication and transport systems highly integrated to enable trade, power concentration, wealth accumulation and inequality. You can't exploit another country, the land or other people, unless you are well connected to them. This is an unethical form of integration. It is these types of global integration that also enable mass pollution, climate change and biodiversity destruction. At a smaller scale these integrated systems are evident in our towns and cities, concentrations of people brought together for security, ***cooperation*** and ***efficiency***. Permaculture shifts us towards a localism and economies that are biassed towards a bioregional level of interaction, a step towards connecting with the land under your feet and the people in your neighbourhood.

Equally if we over-integrate then powers of competition can become dominant and weaker ***elements*** suffer. The planting of a dense forest garden may lead trees to become too tall and thin as they compete for the limited share of light and will cast even more shade to ground level where herbaceous plants will no longer thrive, resulting in reduced ***diversity***. Our desire to integrate completely with the whole world as a global society is simply overwhelming as we become over-connected through the multitude of social media platforms on devices that can access everything, resulting ironically in increased isolation and mental health disorders. The word fascism itself comes from the Latin for a bundle of rods tied together, an enforced ***integration*** and a loss of individual freedom and autonomy.

My preference would be to refer to this principle as ***integration optimised***. We need to ***integrate*** our ***elements*** so they can interact with care and ***cooperation*** with minimised competition. We need to encourage ***diversity***, move away from our desire for simplicity and build supportive, ***resilient*** networks. We need to enable a ***fair share*** of resources to flow and an optimum level of ***integration*** will support that.

And stand together yet not too near together:
For the pillars of the temple stand apart,
And the oak tree and the cypress grow
not in each other's shadow.[9]

Kahlil Eibran

Least Change

Written in collaboration with Marina O'Connell and Steve Charter

Least change for the greatest possible effect

This principle first appeared in permaculture literature in *A Designer's Manual.*[1] It gets an incredibly brief mention as one of the five design principles in the second chapter. It is worded as make the ***least change for the greatest possible effect.*** All Bill writes is 'for example, when choosing a dam site, select the area where you get the most water for the least amount of earth moved'. In other words if you are going to spend a lot of time and energy building a dam (and they sure do take effort to build) then at least put your effort into the design to make sure that the amount of work you do will be minimal and the benefits will be maximised.

Winter pond at Abundant Earth after being dug deeper and wider © Jo Bolton

At Abundant Earth we have constructed several small ponds and one large pond at our smallholding and they have always been in locations where water gathers anyway. We are simply making the ponds bigger and deeper.

I have often thought of this principle as doing the least amount of work required to get the biggest gain. This is not about being lazy though, or about exploiting others to do the work for us. Permaculture designers have an ethical framework that includes people care, and I am yet to meet a lazy permaculture designer. We all seem super busy ploughing our time and energy into creating new ways of relating to the earth and each other. I think Bill's intention with this principle was to encourage us to find creative solutions that involve minimal work and simultaneously create the most beneficial system. We need to work smarter, not harder and avoid burning the candle at both ends to achieve our goals.

Is least change always a small change?

Examples on the Permaculture Association (Britain) website[2] include growing your own salad on a windowsill and insulating your home instead of moving to an insulated house. These are also ***small and slow solutions***, which is exactly where David Holmgren placed the *least change* principle. He clarified this in an email exchange with myself: 'Least Change is clearly part of Small and Slow Solutions in that a design solution by its very essence involves change and that change should be small and slow (i.e. least).'

Another permaculture designer to make a link between these two principles was writer and teacher Kate Martignier, who wrote an article about them on the Permaculture Research Institute's website.[3] She brings them together to focus on exploring the best place of intervention in a system through awareness of leverage points. In particular she explores how making really tiny changes in your life can feel painless and so the change is more likely to be maintained. I'll come back to leverage points later in this chapter.

Many of the least change solutions we find in permaculture could also be defined as small changes. When trying to find the best dates for a group to meet it can be useful to use a pattern, such as always meeting on the first Monday of the month, or always on the 18th of the month or every Friday evening. The appropriate solution will depend on the context. The alternative is the nightmare of arranging every single meeting one at a time and meeting just to arrange meetings. When I was involved in breeding hen chicks, I needed to separate out two sets of mother hens and their chicks as they needed different care strategies. I didn't have a spare hen house but the space they occupied was easily split into two spaces thus minimising construction time and costs. This was especially suitable considering the change was only required for a few weeks until they could be reunited, a change that was both least and small.

For a very significant long-term strategic plan, least change obviously doesn't mean a small amount of work. For example, Evolution Music,[4] cofounded by Steve

Charter, is trying to change the entire music industry through changing the raw materials used to make records. Their new material is a bioplastic, which is starting to replace fossil fuel based vinyl records. Steve focused on the principle of least change and working with nature in the development of the business. Using the existing record production industry's equipment and just changing the raw material to an eco material is an excellent example of least change. They have also been working with the nature of the music industry's involvement in social change and accelerating that change with other allies such as Earth Percent[5] and Music Declares Emergency.[6] Their main work has been refining the bioplastic compound to reach the high quality needed for producing records, and then marketing the product and getting key players on board including Brian Eno and Coldplay amongst many others.

Another example from our smallholding, when it came to rebuilding the compost system in our market garden, I explored the options of a total rebuild in a potentially better location or a repair of the existing system. The repair option turned out to be best and this was supported by the principle of least change. It was still no small task and took several days to complete the project. I think there is a case to keep both the least change principle and ***small and slow*** principle, they overlap but also there are plenty of differences.

Minimum effort for maximum effect

This principle has also been referred to as ***minimal effort for maximum effect***. As far as I can tell this wording was first used in permaculture literature by Graham Bell in *The Permaculture Way* (1992).[7] I think finding the action that is minimal effort can ironically be a hard one to find. Often the easiest solution to a problem is to throw money, time and effort at it but what is required here is careful observation, creativity, lateral thinking, patience and design skills. When we get those moments of genius we find the point of leverage to tip our design in an amazing direction. To find that best minimal effort solution often takes time, allowing the mind to wander intuitively and creatively until that wonderful answer appears, often at 2am, or whilst in the bath or upon waking from a dream. Some permaculture designers suggest we should spend 80% of the time on a project thinking, designing, planning and preparing for the decision, and then only 20% of the effort on the implementation. We are then more likely to use less physical energy in implementing the design change and hopefully it will also be a better solution and last longer. This is a variation of the 80:20 rule, also known as the Pareto principle.[8]

Permaculture designers cannot be given credit for originating minimal effort as a principle. The principle of least effort was first articulated by the Italian philosopher Guillaume Ferrero in 1894 but probably dates even further back.[9] It was further developed by George Kingsley Zipf in *Human Behaviour and the Principle of Least Effort: An Introduction to Human Ecology* (1949).[10] His book is 585 pages and I have to admit I've only read the introduction. Let me know if you do read it and please give me a summary! I do know though that he regarded least effort as a variant of least work.

When seeking information in a research process we will normally stop once we have found the answer that we were looking for, or at least the result that is good enough for now. This is even described on the Wikipedia page for the principle of least effort: 'For example, one might consult a generalist co-worker down the hall rather than a specialist in another building, so long as the generalist's answers were within the threshold of acceptability.'[11] This is a caution though to all of us when searching for an answer. When we get that result that is good enough then it might be a reflection of an unconscious bias. When we find the answer we were looking for then we might stop looking any further. We need to counteract those biases and make more effort to triple check the result.

Pathways of least resistance

The term, path of least resistance,[12] apparently first appeared in the 1820s as an engineering term but has expanded to include human behaviour. I would say there are overlaps between the least change principle and this scientific term. It means 'the easiest course of action, a method of doing something that will cause the least upheaval, opposition, unpleasantness or drama'.[13] We can use this term to describe the flow of water downhill or electricity through cables. We can see many further examples in the behaviour of animals and people because we need to conserve energy for survival. If we make too much of an effort, without the relevant resources, then we will exhaust ourselves and that could be fatal. Throw something

The desire line as path of least resistance © Stephen Andrews

into the air and it will surely drop back down. Heat will move to a cooler location. Wind is air moving from a place of high pressure to low pressure because that is the easiest path for it. Even 'light travels between two given points along the path of shortest time' as stated by Fermat in the 1600s.[14] Put a barrier in someone's way and they will go round it. Look at animal tracks and see how the path is not straight but will bend around immovable objects such as trees. Our natural paths will also work with gradients, gradually going up a hill rather than directly up it. It is recommended to encourage the formation of desire line pathways before the implementation of fixed permanent paving.[15] I love the question that permaculture designer James Taylor uses to find the path of least resistance: "What would this job look like if it was easier?"

Flipping the system

Sometimes we know the functions that we desire from the analytical work we do in a design process. Sometimes we just need to know which elements would be the best choice. For example, we know that we need a water supply to our farm but do we need a dam, rainwater butts, mains water supply or to do nothing and just hope for the rain at the right time. Further analysis of options may reveal which choice fits the *least change* principle.

There are times when we already have a system in place but something is not working and we find ourselves in a spiral of erosion as things go from bad to worse. Finding the best intervention in such a scenario to reverse the decay can be tricky. We need to find the solution that flips the situation towards a spiral of abundance, where things go from not OK to a lot better. Marina O'Connell, founder of the Apricot Centre[16] referred to this as "flipping the system". Ideally the change we make requires little effort. For example, if we are considering taking on a degraded piece of land then it could be a lot of work. We have strategic choices to make that could involve a lot of effort as well as money, time and probably machinery. We might take the middle ground and cultivate the land with deep-rooting green manures that suppress weed populations and slowly increase soil carbon and organic matter. Or we could go with the option of absolute minimal effort by allowing a totally natural regeneration of the land, avoiding all action and ***working with nature*** by allowing nature to take its own course.

A hierarchy of interventions

Change is constantly happening, whether we like it or not, and creating a permaculture system involves change. Working with change can mean identifying the right level of intervention for the given situation, from no intervention at all right through a dramatic revolutionary shift.

Doing nothing could be the best intervention. Leave the land alone to naturally regenerate and wait for the trees to emerge through the process of natural succession.

Doing nothing could be the start of allowing time for the creative solutions to emerge. It could be the opportunity to use the resources already present. It could be working with wildlife and encouraging the principle of everything gardens. Did you know that just imagining the process of exercising your body can increase your muscle strength by 10%? Simply visualising an action, such as getting a basketball in the hoop, can be almost as effective as physically practising a movement.[17]

Maybe a preventative intervention is best, such as maintaining good health with lots of sleep, rest and good food to avoid illness. A stitch in time to mend an item of clothing before that jumper falls to bits completely. Repairing the rabbit-proof fence around your veg garden a bit every winter rather than waiting until it completely collapses and requires a complete rebuild. These solutions often require a bit of time but will save money in the long run. Our annual winter repairs to our market garden rabbit fence often involve spending very little on a few nails and utilising spare parts as compared to the cost of an entire new fence.

Removing obstacles, such as limiting factors, could be the best intervention. If we can see that the change desired is actually being prevented by a particular factor then the best strategy will be to remove that factor. For example, remove the sheep from the field and the trees will eventually appear and you won't have to plant them.

Providing education and training, for ourselves and others, could be the best intervention. A few years ago when tweaking our off-grid solar based ***renewable*** energy system to cope with winters, I had a few choices. I could have spent a lot of money on a bit of kit that would simply switch off the electricity if the demand on the system was too high. Instead I chose to provide a guide and checklist, which I turned into a colourful diagram and put it up in the kitchen for all to see. Then I requested everyone's agreement to give it a go to see if it helped us to conserve electricity, protect the batteries and cope with the low availability of green electricity over the winter months. Over the years we have reduced the length of the time that this guide needs to be followed by increasing ***efficiency*** of appliances and slowly increasing the number of solar panels thus increasing the available power.

Some interventions require spending more time and energy on the process of change but it might be more effective in the long term. If we want to change attitudes towards the planet and enable the survival of life then doing absolutely nothing is not an option. We need to consider which intervention is best in any given situation. When it comes to changing attitudes towards the planet, is it best to remove obstacles to change, provide better education about the planet or make changes in the law? When it comes to the continued collapse of ecological systems, as permaculture designers we want to see maximum change in our societies. And yet here is this principle, asking us to find the intervention that involves the least change. Maybe ***least work*** is a better name for this principle. One of the ways to minimise that work is to identify the best leverage point no matter which type of intervention we choose.

Leverage points

Donella Meadows, scientist and system analyst, devised a twelve-point map of leverage points for intervening in any system.[18] Meadows stated that there are places within a complex system where 'a small shift in one thing can produce big changes in everything'. Donella is most well known for publishing *The Limits to Growth* in 1972[19] and going on to develop systems theory. There is more about Meadows in the ***pattern*** chapter where I explore definitions of systems and in the ***self-regulation*** chapter where there is information on feedback systems.

The twelve leverage points appear in a particular order. At one end we have interventions that are easy to implement but have little to no effect on the system. At the other end we have changes that are very hard to enact but will make a massive shift in the system, changing it out of all recognition. Most of the time we focus on tweaks that are least change and least work but consequently have little change on the overall system, which might be OK if our intention is to maintain the system as it is. When we know a system needs changing dramatically, we need to gather our resources and get to the underlying cause and make the biggest change possible. We may not succeed in making the biggest change but we can try our best. Permaculture designer Colleen Stevenson referred to this as ***use your energy where it can affect the most change.***[20]

Meadows liked to use water and icebergs to simplify and explain her understanding of systems theory.[21] Systems can be compared to a bath or lake, with water flowing in, a body of water stored and excess water flowing out in various locations. She referred to this as stocks and flows. She also used the iceberg as a model, which I think is very helpful as many parts of a system are invisible to us, such as organisational structures, ***patterns***, goals, feedback loops, beliefs and mindsets. This is one of the reasons why we tend to deal only with the easiest interventions as they are also the seen physical parts.

Donella Meadows' Twelve Leverage Points[22]

1. Constants, parameters, numbers
2. The sizes of buffers and other stabilising stocks, relative to their flows
3. The structure of material stocks and flows
4. The lengths of delays, relative to the rate of system change
5. The strength of negative feedback loops
6. The gain around driving positive feedback loops
7. The structure of information flows
8. The rules of the system
9. The power to add, change, evolve, or self-organise system structure
10. The goals of the system
11. The mindset or paradigm out of which the system arises
12. The power to transcend paradigms

So for example, when we consider tackling climate change, we have a wide range of leverage points and an incredibly complex system made up of many many systems. It is consequently really hard for us to understand how to make a change, what change is best and it could take years, even decades before we know if the change made has actually had the desired effect. At one end we could just change who is in charge of an oil company. It's a change but is also likely to maintain business as usual. We could very slowly turn the tap off and gradually lower our carbon emissions from one fossil fuel, let's say oil, but at the same time in another country they are increasing it, or maybe we start using more of a different fuel, like gas. With this approach we still have runaway climate change, just maybe slightly later. At various places in the middle of the hierarchy of leverage points we can consider limiting factors, bottlenecks and feedback loops. The speed and strength of feedback loops is really important, in particular if it's slow, for example you make a change but it takes ages before you know if anything is different. Ideally we need agile systems that are quick to respond and adapt so we can see the result of our intervention quickly and know if we have done the best thing.

Further down the iceberg we find leverage points that start to become more effective but even harder to do, such as changes in access to information and changing the rules. For example, exposing the oil companies that funded climate denial information meant that they lost most of their respect and trust. Knowing which type of transport has the lowest carbon footprint enables us to make a choice. When Gorbachev came to power in the Soviet Union he changed the information flows (known as glasnost) and changed the economic rules (called perestroika) and that led to the collapse of communism in Russia and Eastern Europe. Unfortunately that didn't stop the rise of another dictator. We need laws and rules to help us combat climate change and they are happening bit by bit but they are not enough.

At the far end of the leverage scale there are shifts in values and mindsets. This could mean we prioritise and place the highest value in nature. By having this different belief about nature, we will in turn change our goals, priorities, rules and feedback systems. This would be a paradigm shift. Meadows' theory is a hard one to fully understand but is based on her decades of practical application and analysis before her death in 2001. I encourage you to explore it further if it interests you and see how it could support you in making effective changes with the least amount of effort, even if that does include some hard work. I believe there is a possibility that the permaculture principles can contribute to the paradigm shift that we need. Once we understand them and start to apply them more consciously, we start to see our goals and behaviour change too.

Limiting Factors

Written in collaboration with Sam Woods, Jo Holleran, Cathrine Dolleris and Chris Evans

Design tool or principle?

Most permaculture literature and online sources refer to *limiting factors* as a design tool and not as a principle, furthermore in most cases it is entirely focused on the ***removal of limiting factors***. We can see this on the Permaculture Association website[1] and in Aranya's book.[2] There are very few books that explore limiting factors in depth but they appeared early in the teaching of permaculture as Chris Evans has described to me: "I remember on my PDC in 1989 in New Zealand, Bill Mollison and Lea Harrison talking about limiting factors and also calling them 'building blocks'. I watched Lea continue to deliver that session on courses that I shadowed with her in Nepal and the UK. I've used the concept ever since in my own training. Understanding them is a key part of observation."

Chris Evans, co-author of the *Farmer's Handbook* (2001),[3] describes limiting factors as "building blocks that can both limit and aid the progress of a design". Factors are divided into two sets, one visible and the other invisible. For example, to plan and understand the life cycle of a crop it is essential to understand all of the things that can go wrong and adapt your strategies accordingly. The invisible limiting factors are just as important and often harder to overcome and plan for due to their hidden nature.

The gradual chipping away of these limiting factors can be a life's work and it may involve many of the other principles to create effective strategies, such as ***efficiency*** measures, ***catching and storing energy*** and accurate ***observations***.

The other key source of the limiting factors concept for permaculture designers appears to be from Ian McHarg's book *Design with Nature* (1969),[4] and the first permaculture designer that I can find that mentions him in the UK is Mark Fisher, who was one of my permaculture Diploma Tutors. He provided me with a handout back in 2002 that included details about McHarg and the identification of limiting factors. That handout is still available on Fisher's website.[5] More about McHarg later on.

The most common place to consider limiting factors is in the observation, analysis and pre-decision making phases of a design process but they can appear

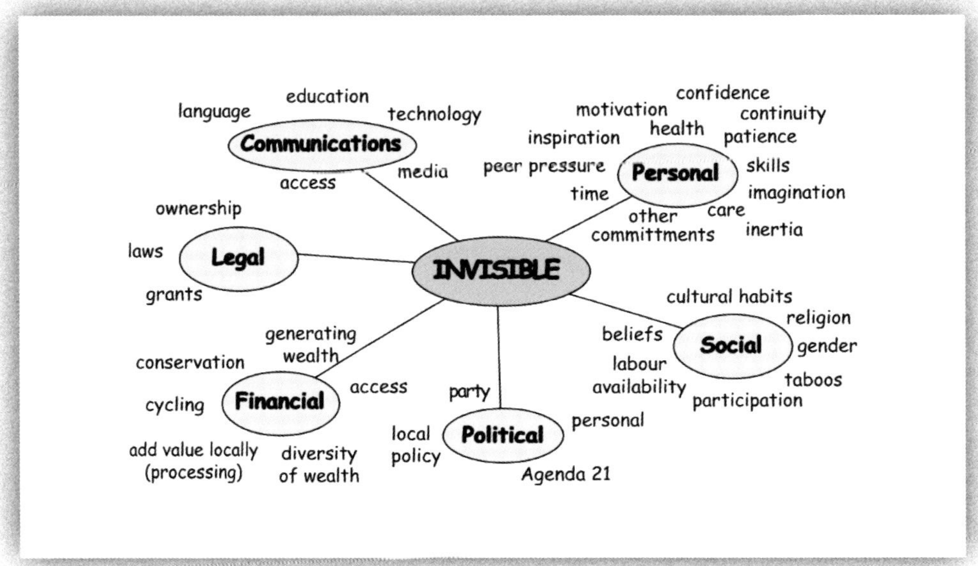

Invisible limiting factors map © Chris Evans

at any stage. I frequently use them in the development and tweaking of a project rather than at the beginning. Typically we want to save ***energy*** and ***work with nature***, and so if we can simply remove what is holding our plans back rather than fighting against the barrier then we may have a low-energy solution. This could either be the unblocking of a natural energy flow or the plugging of an energy leak. For example, when I decided it was time to expand our veg box scheme at Abundant Earth, I focused on the factors that were holding it back rather than how I could make it expand. Those limitations included space for packing bags and the number of pick-up points. By changing the internal layout of the packing space (a simple ***least change*** removal of a low and fairly redundant shelf) and the addition of an extra pick-up point meant that we could add ten bags to the system. We already had a waiting list so bringing on new customers was easy. I regarded my use of *limiting factors* in this context in the same way as I would use other principles i.e. as a thinking tool, focal point and guide. I currently do not want to expand the veg box any further as that would take it beyond my capacity. If I did expand the scheme too quickly without planning the next stage of development, then the quality of the service could drop, mistakes might start to appear and I may feel overworked and all of that would be hard to sustain. I like the current limiting factors in place and I know exactly what they are. If and when I am ready then I may consider removing some more limitations in the future and expand the scheme again.

I think by only focusing on one aspect, i.e. the *removal* of limiting factors, we are missing a trick. I can see why it is regarded as a design tool but I have always

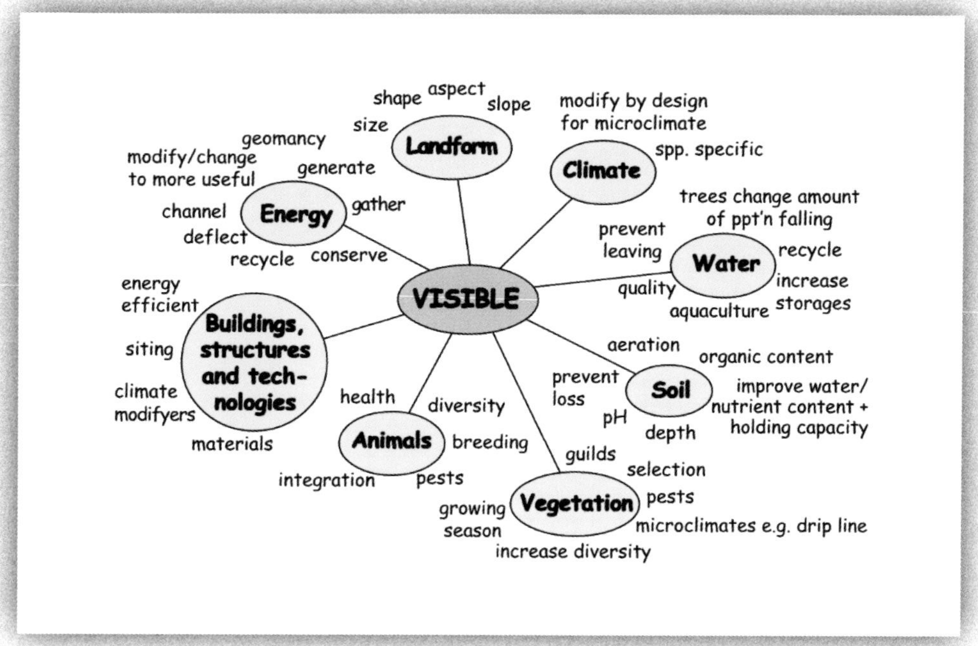

Visible limiting factors map © Chris Evans

seen it as a principle as well. I don't always want growth and development of a system. We need to be aware of limiting factors because we might want to keep them as useful or we might want to remove them as part of our design solution. Limiting factors have pros and cons based on the context. Some of the other principles also sit in this grey area of being both design tool and principle, such as ***observation*** which is part of a principle as well as being part of various design tools and a key phase of design processes. I have already shown how some principles can also be overplayed and underplayed e.g. ***integration***. My proposal is to put forward ***limiting factors*** as both a design tool and as a principle.

The leaky barrel

Historically there have been several sources and developments of limiting factor theories and they all predate permaculture.[6] A common exploration has been in the biological processes of plants where the limiting factor to growth could be the limited access to a key nutrient. This has been really important for the understanding of crop science, agricultural developments and the development of artificial fertilisers. At a broader scale, ecologists also consider limiting factors such as food supply, energy, space and predation in population dynamics, for example the population of rabbits might be limited by the number of foxes in the same area. Together these population limits are the checks and balances on a system and create a carrying capacity for a species in a given location.[7] There are varying figures about the carrying capacity of humans on planet Earth but it looks like we have overshot all of them.

The concept of limiting factors is largely based on Liebig's Law of the Minimum (which was actually developed by Carl Sprengel in 1840), which states that growth is controlled not by the total amount of resources available, but by the scarcest resource.[8] We can visualise this concept as a barrel with staves of unequal length. The amount of water in the barrel is limited by the lowest stave, in the same way that a plant's growth is limited by the nutrient in shortest supply. We could also use an image of the weakest link in a chain or of a dam holding back a reservoir. The notion that there is only ever one limiting factor at any given point in time and space is important to know. There may be other factors waiting to become limiting factors but until they are given the top job, they aren't limiting factors. Once the key limiting factor is removed it will be replaced by another limiting factor further down the line (the next highest stave if we consider the image of the barrel again). Another key idea in biology is that limiting factors can trigger competitive behaviour. If the limiting factor is key to survival, such as food availability or space, then behaviours may change towards more competitive approaches. This indicates that ***cooperation*** is more likely between individuals when key survival based limiting factors are minimised. If we all have access to good food and a fine roof over our heads, our core needs are more likely to be met and we may be nicer to our neighbours.

Using limiting factors and ecological succession in land management

Want to create a woodland? Ask yourself why this land is not woodland already. Maybe there is access to the land from grazing animals, either deliberate livestock or wild herbivores. By minimising the herbivores, as the key limiting factor, we can unblock the natural process of ***ecological succession*** and a woodland may appear with time. We can ***accelerate succession*** by removing limiting factors or it can be held back by keeping limiting factors in place. For example, if we stop ploughing the land then grassland will emerge. Equally if we want to keep land as grassland then we need to keep herbivores on the land or have a seasonal cutting regime. Conservationists want nutrient levels to be limited in a species rich wildflower meadow.[9] The annual removal of grass from a hay meadow keeps nutrient levels low in the soil and this in turn favours species that can thrive in those limiting conditions, which tend to be rarer species. In a forest garden setting we may want to increase light to the woodland floor or we may want to encourage shade tolerant plants at ground level, either way we will want to manage the branches that are limiting light levels by either leaving them alone, minimal pruning or extensive loping. These are all examples where we need awareness of the limiting factors and know which ones to keep and which ones to remove.

Limiting factors in food crop systems

Typical limiting factors in a food crop system could include light, temperature, soil type, soil nutrient availability, water, aspect, altitude, wind and of course the amount of space available to grow crops. It is important to know which of these is most relevant in your situation so that you can develop appropriate solutions.[10] There is no point planting light-hungry plants in a shady location or crops that need lots of water in a drought area unless you are going to develop strategies to counteract these situations. If we know that temperature is the limiting factor then maybe we need to consider changing the microclimate of the growing area with, for example, a windbreak or addition of a greenhouse, or we need to stick to plants that tolerate the temperature range.

Soil type is also key to deciding which crops to focus on. At our site near Durham we have a clay type of soil which is not suited to the growing of carrots (they prefer sandy soil) so we do not bother growing them here. We are also limited by the size of the land and what areas within that plot are suitable to grow crops. For example, a lot of our land is mature woodland, not suitable for growing most food crops but it is great for wildlife, firewood and craft activities. Knowing the limitations of your particular site is why you should not copy someone else's strategy without first understanding the reasons for their strategy. On one site adding drainage might be appropriate, at another site the opposite might be required.

Involving volunteers is a common strategy to overcome the limitation of time and energy in food growing systems. This needs to be adopted with caution though as volunteers can cost you time and energy in setting up a task, training in how to do the task and their competence in the ability to do it to the standard

The extra large storage bay at the end of the compost system © Wilf Richards

that you require. Be mindful not to replace one limiting factor with another that is equally powerful. When we identify a limiting factor, we may consider other principles to support an appropriate strategy. For example, if the limitation is water supply then ***catch and store*** rainwater, or if the limitation is time and energy then develop your site in a ***small and slow*** way. Limiting factors can change over time and as one limiting factor is resolved another may become more apparent.

Overcoming a limitation in compost production

In my 2021 compost system rebuild design,[11] I identified a limiting factor in the production of compost and overcame that by rebuilding the last bay as a storage bay five times the size (see previous page). There was plenty of compostable material coming in but turning into the next bay was restricted by the speed of use. This speed of use varied during the seasons. In spring and summer the use of the compost is at its fastest as new beds are made and crops planted out. But in autumn this process slows down and in winter it almost grinds to a halt. There are times of the year when there is plenty of weeding and material to harvest to make compost but the end use is slow. If the end bay was full and compost use was low then this could create a bottleneck[12] in the system. I could not turn the compost if the end storage bay was full. By making the end bay a lot bigger I could continue the cycle of turning the compost all year round. Annual production of compost doubled after this construction change from around five to ten cubic metres, thus adding to the principle of ***yield unlimited*** too.

Limiting factors that determine placement

So far we have mainly looked at limiting factors that hold development or growth at a particular point. Ian McHarg's book[13] looked at another aspect of limiting factors which was related to placement. McHarg was well ahead of his time and has been credited with bringing ecological planning methods into the mainstream of landscape architecture. His approach was to create maps and overlays, adding positive opportunities and negative constraints to a site. The positive opportunities included ecological options such as wetlands, as well as human needs, such as housing. The negative constraints were the limiting factors for placement e.g too rocky to farm the land, too steep for a road etc. Through careful questions exploring opportunities and constraints, then adding these as map overlays to a site, the result was some great design decisions. McHarg added to the typical design question of "Where can this go?" and asked as well "Where can it not go and why?".[14]

In the placement of a new road we can identify all of the limitations on where the new road should *not* go e.g. on top of existing housing or through protected wildlife areas. By marking on the map the areas where the road should not go we can identify the limitations of placement for locating the road. If we want to consider the location of a new house that is gravity fed by the spring water nearby, then all locations uphill from the spring must be ruled out. McHarg's method, also known by permaculture designers as the *McHarg Exclusion Method*,[15] became the basis for Geographical Information Systems.[16]

Limits and personal progression

In Looby Macnamara's Design Web one of the key anchor points is *Limits*. Here she asks us to identify what is blocking our paths, what might keep things ***small or slow***, what is holding us back, what our limiting factors are and why we would not want to change. Looby's book *People & Permaculture*,[17] which includes details about the Design Web, focuses strongly on personal, social and community development. It includes a section about the invisible as well as the visible limiting factors (pp90-91). Those invisible factors highlighted include overcoming the barriers associated with confidence, skills, momentum and many more. Inevitably this means the focus is on removing limiting factors, which is appropriate. The process of identifying our limits can be balanced by identifying what can help and how we can flip the limits into goals and design functions.

We can all have many personal limitations and barriers that we wish to overcome. They could be the lack of a skill, lack of key knowledge or maybe a self-limiting belief. It could be time and money, the two classic limiting factors to anyone and anything. There may also be poor communication skills, a lack of motivation or the energy to carry out a plan. These limitations could be physically real or more a matter of perception and belief. Identifying these types of limitations that are holding us back and then finding ways to remove them could really increase our ***yield***. There are many tools we can use to work through such invisible factors, one of which could be the *Reframing the Belief* tool used by Think like a Tree[18] facilitators. Another could be lateral thinking tools and finding ways around an issue. Maybe we can gain more time through increasing the number of times we decline an offer or have more money by not spending as much rather than trying to earn more.

We may also need to be kind to ourselves and simply accept our personal limiting factors. Some things are not in our control. Some things might change with time and some things won't. Patience might be required to wait for a timely moment to push for change. As the old saying goes 'no flower blooms all year round'. Maybe we do not have the capacity to push back or develop that new skill at the moment. Maybe we just don't have the ability to access that extra time or funds that we need right now. We all have limitations and many of us have a range of disabilities. We may need to accept what we can't change in our lives. Personally my vision in my right eye is a total blur and no glasses can help. Some situations that may bring us grief could also be design opportunities to explore those personal limitations and find coping mechanisms. We all have an ultimate limitation, as Sarah Spencer pointed out in *Think like a Tree: The natural principles guide to life* (2019),[19] a principle we can't ignore: ***know that death is part of life***.

Limiting factors can be positive

Some limiting factors I believe are good for us. Maybe we want to keep certain boundaries and limiting factors in place, such as ***self-regulation*** of our behaviour. This may include boundaries on how we want to work with others, what we are willing to accept and compromise on, how assertive we need to be or maybe a discipline you wish to maintain such as limits on junk food, alcohol or social media. These may be strategies that we adopt to avoid burn-out, maintain rest or lower stress levels. There may be certain elements in our systems that we want to limit but not remove, like a bug that eats a food crop but is also food for a friendly bird. By limiting the bug we will increase our crop success but may also limit the predator of that bug. Wipe out that bug with pesticide and you will also limit the animal that fed on that bug.

The permaculture ethic of ***fair shares*** asks us to put limits in place. We need to acknowledge that we live on only one planet and thus need to welcome limitations on our consumption and behaviour to live within the means of a single planet. This might include no harm to animals, minimising flying, only using organic methods, always buying the most environmentally friendly item that you can afford and protecting wildlife areas from unnecessary harmful development. We need to place limiting factors on corporate and governmental behaviours that are harmful to people and the planet.

The principle of ***small and slow*** solutions can provide an opposite to the removal of limiting factors, although as you will read in that chapter we do not always want to be small and slow. The aim of removing limiting factors can imply we want things to get bigger and go faster. A balance is to be struck here between these two intentions. There are times when we need to go slowly, when we want to procrastinate and wait for the time to be right or maybe when we need to rest and recover. There are times when we need things to be small, and there are times when making something bigger is completely pointless and paradoxical. This is most neatly presented as Parkinson's Law which was originally focused around the amount of work carried out by civil servants and the observation that the amount of work to be done increased when the number of workers increased.[20] This sounds sort of logical but if a task has a limited size, then you would think that by putting more people on the job it would be less work per person and the job would get done faster. The opposite can occur. It turns out that work can expand to fit the time provided. Parkinson's Law can also apply to stuff. Stuff expands to fit the space provided as the amount of stuff you store is based on the amount of space you provide for storing stuff. It also applies to roads and cars. The more roads there are, the more cars there are on the roads and the more time is spent in cars.[21]

Limitations can give us structure and precision. Trying to complete a project without a deadline can be pretty difficult, having a time constraint can provide just the right amount of pressure to get a job done. And it is really nice to know that some things will come to an end and not go on and on and on. When we work in our gardens we are aware of the seasonal changes and the limitations that each season has. We need to be timely when we sow seeds and harvest just at the right time as we ***work with nature***. Schools and workplaces can provide us with schedules, deadlines, rules and structure, some of which can be deeply unhealthy and dictatorial but done with care they can be super useful. Boundaries can keep us safe and ironically provide freedom within the ***edges*** of that boundary. Limiting our choices, either by using ethics or through necessity, can provide a speed of decision making and lessen our anxiety and overwhelm of options. We can just choose what is good enough for now.

Limitations can bring us expertise. Johann Wolfgang von Goethe wrote in his poem about nature and art, *Natur und Kunst*, in 1802, that mastery is revealed in limitation: 'In der Beschränkung zeigt sich erst der Meister.'[22] By knowing the rules and limitations that we are restricted by, we can develop a skill within that framework. If we want to broaden our focus and learn many skills then we will not master any of them, we will become a 'jack of all trades' instead. This may limit our mastery but could be beneficial for flexibility, as described in the ***adaptation*** chapter.

Limitations can bring us creativity. Give yourself a problem to solve with limitations on time and resources and you may find the perfect bodge solution that will not only do for now but might even be an act of genius to others. In 1975, musician Brian Eno and multimedia artist Peter Schmidt published a deck of cards called Oblique Strategies.[23] Each card provides a statement designed to provide a limitation upon the artist with the intention of making a creative breakthrough through the use of lateral thinking. They are also available as a free app and I use them on an occasional basis with the band I'm in, Queen's Screech.[24]

Is the sky the limit? No it isn't © Wilf Richards

Observe and Interact

Written in collaboration with Lusi Alderslowe

Can you see what I see?

Observation is a key tool in the permaculture approach. In fact it is a key tool in all approaches and disciplines because what we observe is not the whole story and there is always more to observe. Our place in this world and the space around us is constantly changing and always unique. We can add to that our brain's ability to filter our observations, our physical capacities and limitations, and the position from which we make our observations. This position could be physical e.g. close up or far away, or it could be how our prior experiences, expectations or bias influence our observations.

The specific nature of each situation we find ourselves in means we need to be careful with our observations. It would be foolish to take a successfully implemented plan and exactly replicate it in a new location. Instead we need to be inspired and see what aspects, such as principles or approach, can be copied into the location we are working in. Observations are key to knowing what is already in place, what is missing, what is possible, what is improbable and what is impossible in any context. Sometimes our observations are not clear and we can carry out tests and research to learn more. These interactions with our observations enable us to test the truth of what we perceive. As the world is constantly changing, this can be something we need to repeat. We can ask such questions as: Is this still true? Have I got this correct or good enough? Does my understanding of the world still work? What have I learnt? What do I think happened here? This open and curiosity-based approach to observation is crucial. We can see in nature animals and plants observing and interacting with the world around themselves, whether that be following a scent, the subtle daily movement of plant leaves to track the sun or pushing down roots towards the moist soil. Equally, as gardeners, we observe and interact with our crops, trying to learn and know when that plant needs water or better soil.

The importance of observation was expressed right from the first permaculture books, focusing to begin with on the importance of observing ***patterns*** in nature, climate, soil, landform, and natural habitats. Techniques and methods of observation were shared in *A Designers Manual*[1] such as thematic, instrumental and experiential methods. Despite the clear importance of observation, as included in most permaculture books, it was not elevated to the status of a principle until David Holmgren added it to the mix as one of his twelve, wording it as ***observe and interact.***[2] The basis of design is to start with observations and through processes of analysis and careful interaction we can find ways to minimise work, limit unnecessary energy inputs, reduce pollution all whilst increasing the harvest. Holmgren tied into this principle the proverb *beauty is in the eye of the beholder* to highlight the need for unbiased observations free of our judgements and values, or to at least openly acknowledge and be aware of when we are making observations tainted by evaluation. 'Ethics and ideology act as filters that determine what and how we see. These filters are unavoidable – in fact, essential – but the rush to judgement of right and wrong frequently clouds our observations and prevents understanding.'[3] Holmgren was influenced by the early attitudinal principles when devising the details of this principle, such as ***least change***, ***working with nature*** and the ***problem is the solution***.

Use thoughtful and protracted observation of natural systems rather than protracted and thoughtless labour.

Bill Mollison[4]

How can our senses help?

There are many approaches to observing this world around us. We obviously have all of our five main senses for a start. Not just eyesight but broader observations through hearing, touch, taste and smell are all important. And our senses are not just limited to those five, depending on what research you read, it is now recognised that we have many more senses, such as spatial awareness, balance, temperature and the passing of time.[5] Many plants and animals have incredible senses too, many far better than our own, such as a dog's sense of smell or a plant detecting a chemical signal in the soil. Furthermore some animals have senses that we do not possess, such as bats picking up magnetic fields.

Our senses work together to create an understanding of the world around us. In the *Earth Care, People Care and Fair Share in Education, Children in Permaculture Manual* (2018)[6] the senses described above are called 'eyes' and are included alongside head, heart and hands to ensure that kids have a 'rich multidimensional learning experience… designing sessions to engage the whole body'.

When we identify plants, the texture or smell of a leaf can be more of a clue than the shape of it at times. In fact it is essential to use more than one of our senses when it comes to identifying plants. For example, at first glance the leaves of alder and hazel might appear similar and they can both grow in woodlands but

Which one is alder and which one is hazel? © Wilf Richards

Is this a mint? © Wilf Richards

hazel leaves are softly hairy. Our awareness can also extend beyond those physical senses too. We may also include our gut instincts as an observational tool but we need to be careful and not use it on its own as it will provide us with observations mixed with prior experience or emotional layers.

The plant in the image opposite could easily be a type of mint but if you could smell it then that would help a lot as it smells of lemons and is in fact lemon balm (*Melissa officinalis*). Just to add to our confusion, lemon balm is part of the same family as mint called *Lamiaceae.*[7]

We can also build upon and use the observations recorded by other people but it helps if we can also have direct experience to help verify others' observations, as their observations might not be accurate either. We live in an age of an abundance of information that is very accessible, whether that be through formal education or the internet, but as we know it's not all correct!

Reading the landscape

There are many ways in which we can deliberately focus our observational skills. We could, for example, do a *thematic observation* where we focus on a topic and look around for just that one thing across a site and create a map. This could be listening for a particular seasonal bird or hunting for a certain spring flower, or the mapping of the ash trees on your site which are suffering from dieback. This method usually involves having a good walkabout. Alternatively we could use the *experiential observation* technique by finding a good place to sit and observe everything we can from that one location. That might include observing the sun and shade, the wind and shelter, plants, insects, birds, mammals, people, buildings and paths etc. With time and experience we may piece together the jigsaw of our observations and start to know the ***patterns*** of our landscapes. Patrick Whitefield referred to this as *reading the landscape.* I highly recommend Patrick's original book *The Living Landscape: How to Read and Understand It* (2009)[8] which covers his extraordinary observational skills. There are many things we can observe in the landscape including the species present, the wind, sunshine, water, land shape and soil. Sometimes we can sense these things directly and in other cases we may need instruments, technology, colleagues, identification books or even experiments to understand more about what we are observing.

There are many instruments we can use to extend our observational powers. For example, devices to measure temperature, wind speed, soil pH and the amount of rainfall. We can use technology to help us with observations too, whether that be wildlife cameras for night time animal sightings or time-lapse cameras for those subtle sun-influenced movements of plants during daylight hours. There are also many apps and websites that help, such as those that show sunrise, sunset and sun height for any location on the planet and at any time of the year. We may wish to also try some observational experiments, such as trying two different methods of growing a crop in the same bed.

A good way to organise the things we observe is to use a mapping tool called PASTE, which stands for plants, animals, structures, technology/tools and events. Another possibility is to create a base map of your site and then add layers on top of the map showing a particular strand of data, such as sunny spots and frost pockets. Once we have gathered our observations then we may begin to analyse the information and see ***patterns*** such as seasonal markers, prevailing winds, microclimate variations, windiest month and driest month etc., all of which are opportunities to ***cooperate*** with the landscape. Observations are ongoing and we may need to repeat them over many years to check our results or see what changes. Looking out for extremes can be particularly useful. I can recall my joy when on a teaching day, focused on observational skills, it was very very rainy. The participants didn't seem so happy when I suggested this was a great opportunity to see the flow of water across the site, looking at drains, ditches, gutters, puddles and leaks etc. Fortunately everyone did get motivated and had good waterproof gear, which resulted in identifying design opportunities and challenges. We can apply observational powers in social projects too but use different methods, such as surveys and focused conversations.

We can also just simply observe for the joy of observation. We don't necessarily need to be consciously interacting with or recording our observations. We can observe and feel awe, gratitude and wonder for the world around us. I call this *seeing the magic*. It's a principle I like to include in my designs especially when gratitude and pause are needed. The best way to do this is using meditation and mindfulness techniques to sit still, take it all in and actively resist the desire to deliberately interact or take notes. This could be what Colleen Stevenson was referring to in her principle: ***slow down and observe***.[9] It's good to stop being so focused on our own thoughts, and instead take a moment, look around us and give something else or someone else in the world a moment of our attention.

The limitations of our observational powers and the need to share our observations

As humans we have physical limitations on what we can observe and how we can observe. For example, our eyes are on the front of our faces unlike most birds, where the eyes may be on the side of the head and consequently they find it a lot easier to see what is behind them. Also we only perceive a certain range of lightwaves. We can't see ultraviolet light like bees or infrared light like mosquitos and some types of snakes. Some of us even have brains wired in such a way that the senses can get mixed up, for example in the condition synaesthesia, where sounds may appear as colours or taste as a sound.[10]

There is also a whole cultural baggage that can influence how we observe the world. Our brains will filter information partially physically and partially culturally. Confirmation bias, which we all experience in some way, shows that when we are presented with new information, what we take on board is massively influenced by our existing beliefs.[11] The risk is that we only see what we believe in already. The art of the researcher is to enjoy being proven wrong, embrace failure and move onto lessons learnt.

Our brains may see patterns where there aren't any and may unscramble things to make them easier for us to understand. The following text can be read by most English speakers, even those with dyslexia. That is because when we read a word it is thought that we do not look at individual letters but instead look at the whole word and various clues such as first and last letter, double letters and pairs of letters that make certain sounds.

Do yuo fnid tihs smilpe to raed?
Bceuase of the phaonmneal power of the hmuan mnid, msot plepoe do.[12]

When we share our observations of the world we may build a connection and trust with someone else if they see things the same way that we do. There can be curiosity, learning and joy in realising that we see things a little bit differently or someone draws our attention to something we previously were unaware of. We may see and possibly counteract cultural and personal bias when we share as that may be a reason for those differing opinions about what is being seen. There can of course also be terrible conflict when our perceptions of the world clash so we need to also apply ***cooperation*** and ***people care*** when we share our observations.

Our process for learning can be through what we observe others are doing. We can help others to learn by sharing our processes and slowing down our actions to make it easier to observe and copy. Eventually we may become fluent in a new skill through this process of observational exchange. When I was involved in breeding chicks in our hen area I was frequently delighted and amazed to see mother hens clearly demonstrating various skills, such as feeding and cleaning,

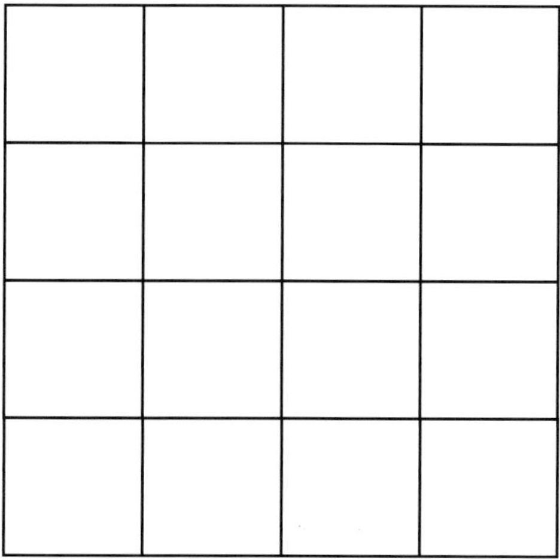

How many squares are there?

to their chicks. We may also need to share our observations when we can see a disaster on its way and provide a warning to others around us. This might involve checking with others to see if they too see the same danger. The same might apply in seeing something that breaks our unwritten cultural agreements, such as rude behaviour. I have seen hens put other hens in their place when their pecking order is forgotten by the younger birds, the elders telling the young ones what is, as far as they are concerned, right and wrong.

I love asking how many squares there are in the image above, as the answer looks obvious to begin with, as sixteen connected squares but there are a range of responses I receive, usually somewhere between sixteen and forty. The point is not the actual answer but the variety of ways that we see the world and it is great to share our perspectives to see what we may have missed. If you do want to know the mathematical answer to this puzzle then see the footnotes.[13]

Slipping into judgement and observation without evaluation

An important trap for observing is judgement. I have already established that due to our brain filtering information and limited observational powers, we can be sure that what we observe is not going to be the whole truth. It is difficult but important to try to be unbiased and non-judgemental in our observations, at least to begin with. And when we do add a judgement we can acknowledge it as just an opinion or a possibility.

One of the key principles from non-violent communication, also known as NVC, is *observation without evaluation.*[14] Making observations in this way can be compared to describing an event as if it were made by a video camera. For example, it is pretty clear which of the following statements is an observation without evaluation and which one is an observation with evaluation.

"Bill hasn't replied to my last two emails about the event."

"Bill never replies to my emails, he is so lazy."

These judgements or evaluations may be based upon notions of good and bad, right or wrong, likes and dislikes. When we are using a design process, which typically begins with a survey or observations, the information we gather should be free of judgements. Just try to gather facts rather than adding opinions in. For example "I can see four mole hills in this area of the garden" as compared to "Those darn moles have been ruining the lawn again". If you are working for a client then one of those layers of facts might be the client's opinions. You might observe that the client thinks their garden is awful but you can leave your own judgement until after you have carried out your observations.

Getting the attitude right

Many with access to land are keen to jump in straight away with ideas and plans of what they intend to do with it. Ask instead what the land wants and needs, not just what you want. Observe it and ask, "What is this land saying to me?". In this process you might see indicator plants showing soil type, pH, compaction and water content in the landscape – all good clues as to whether or not your ideas might actually fit this location or what interventions might be best. There may also be the need to ask for permission and support from the local community. You might own the land but who else is a stakeholder in the land? Who else knows its history? This is especially important in community projects and locations where local and Indigenous people must be more involved. The lack of respect for those with traditional ecological knowledge led to the proposal of ***principle 0*** (zero) by Sarah Queblatin.

> *Honor traditional ecological knowledge or Indigenous wisdom that are already held within a place. Acknowledge and involve its stewards – visible and/or invisible; past and/or present.*
>
> **Sarah Queblatin**[15]

It is called principle zero as it should be considered before Holmgren's principle of ***observe and interact***, which is the first principle[16] in his set. This was articulated in *Permaculture* magazine[17] and in a YouTube interview with Rosemary Morrow.[18] The premise of it is to decolonise permaculture and ensure that Indigenous people, spirits of the land and ancestors are involved in decision making or seeking consent where relevant and possible. To me it feels like it sits somewhere in between the ethics and the principles, or within zone 0, being focused on a deeper respect and understanding for the history and traumas of Indigenous peoples and land. There is a risk that we step into the principle of ***observe and interact*** focused solely on our own needs, even though that shouldn't be the case when we consider the ethics as well. But we have blind spots and cultural bias so ***principle 0*** will help us remember, especially where we work with those people who still have traditional ecological knowledge. Maddy Harland, editor of *Permaculture* magazine shared an example of how she uses the principle: "Sarah's principle not only engages with the land which includes the Indigenous people's past and present but it also asks the land what it wants. Dare I say, I try to do this in the woods. I wouldn't for example immediately go in and fell a tree or coppice an old hazel stool. I explain what I am going to do and why and ask if that is OK. There are times when it is not or I have to wait. But maybe that is a little too radical for many readers. However, we talk about the interconnect and sentience of all things but we rarely act as if this is so."

Another way of tweaking our attitude around this principle has been explored by permaculture teacher Liz Postlethwaite.[19] She asks us to consider what the difference would be if we chose to ***observe and interact*** *within* rather than *with.* This profound shift in language immediately integrates us into whatever we are observing rather than treating our subject matter as separate from us.

I recently read a design by architect and permaculture designer, Edward Dale-Harris, about a house construction in Rwanda.[20] The implementation involved Indigenous adobe brick production that required the makers to observe and interact with both materials and weather. Our industrial mindset may want to know absolutes, such as exactly how long the bricks should be in the sun and the ratio of water to clay. The best approach though is not an off-the-shelf recipe, instead it is a look-and-feel approach from years of experience and ongoing day to day observations. Each batch of soil might be slightly different with variations in clay and sand content, the strength of the sun will vary as will the rain. Compare this hands-on approach to trying to learn from a book. Maybe it's time to put this book down for a while and go and play outside.

Patterns and Details

Written in collaboration with Keira Oliver, Ed Tyler, Tomas Remiarz and Tom Henfrey

This one could spiral out of control

The first in-depth exploration of patterns in permaculture literature appears in Bill Mollison's *A Designer's Manual*[1] which he continues with Reny Mia Slay in *Introduction to Permaculture.*[2] Both contain entire chapters focused on pattern understanding but at this stage it was not proposed as a principle but as an underlying method. When we explore the notion of ***working with nature*** and try to understand the nature of nature then we quickly encounter patterns and there are so many of them. Patterns became a principle when David Holmgren revised the principles at the turn of the millenium and named it ***from patterns to details.***[3] Patterns feature in most permaculture books as a core foundation. More recently Heather Jo Flores named this principle ***recognise and respond to natural patterns,***[4] Starhawk referred to it as ***recognise and work with patterns***[5] and Sarah Spencer had two connected principles:[6] ***follow nature's patterns*** and ***view the big picture then the detail.***

Patterns are included as one of the anchor points in Looby Macnamara's Design Web[7] too, a stage where we are asked to identify the helpful and unhelpful patterns in our design subject and to consider where these may become spirals of erosion or spirals of abundance. We can even see the term pattern appearing in multiple definitions of permaculture, including one by David Holmgren: 'permaculture is consciously designed landscapes which mimic the patterns and relationships found in nature'.[8] It would be easy for this chapter to spiral out of control or branch off in many different directions as patterns are a massive subject and could occupy an entire permaculture book. I am sure we can all be overwhelmed and obsessed by patterns, as Holmgren pointed out and as the old saying goes, *can't see the wood for the trees.*[9] Let's try to break this down step by step and maybe use one of Myron Rogers' systems thinking maxims 'start anywhere, follow it everywhere'.[10]

Can't see the wood for the trees © Wilf Richards

Repeat, repeat, repeat: what is a pattern?

Patterns come in many forms and usually involve a repetition of shapes, sometimes in a predictable manner, and are usually visually pleasing. They are not just a visual form of art. We can see them in geometry, maths, behaviour, relationships, communication, music, buildings… Let's face it, they are everywhere.

When we say we can see a pattern we are saying that we have seen something similar before. It might have been at a different scale, in a different location, in a completely different context but it feels familiar. We recognise it as something that already exists. Our ***observational*** skills are key to recognising patterns so that we may ***work with nature***.

The most obvious patterns that most of us will think of first are the visual pattern shapes such as spirals, waves, weaves, radial, geometric, grouping, branching and overlapping. We can see examples of each of these in natural settings, and each one has characteristics (functions) and applications. Patterns can occur at various scales, for example, branching can be seen in the tributaries of a river, in the veins and arteries of our blood vessels and of course in the form of a tree. The characteristics of each pattern are important. Once we understand these then it can give us an indication of how and where to use them. For example, if we are seeking strength then a weaving pattern, such as we see in a bird's nest and a basket, may be best. Whereas if we are seeking efficiency, such as minimal use of materials in a building, then a geometrical pattern, such as a geodome, might be best.

Beyond the obvious pattern shapes, we begin to go sideways into different types of patterns that are more abstract and harder to visualise. There are patterns of behaviour, such as movement and body language. We recognise body language as a way of communicating the internal feelings of the other person and have to work out if our judgement of the movement we see is correct. Many patterns of behaviour are determined or influenced by environmental factors. Birds, eels, whales, turtles and various other animals migrate in response to seasonal changes, such as temperature, light levels or food availability. We can see the effect of sunrise and sunset on our daily patterns of behaviour, such as sleep and mood, as light levels influence our hormones, including melatonin and cortisol. Fish and plankton will move daily between deeper water and shallower water depending on the time of day, light levels and movement of water.

There are patterns of relationships, which could be defined by the flow of energy in the relationship. For example, parasites that flow in only one direction compared to flowing in both directions between two cooperating species. The pattern in a relationship may be more to do with the phase in the life of the individuals – a mother wants to nurture her son and ask him questions to show care but the son wants independence, autonomy and to be left alone.[11] Those phases of life are a pattern: birth, puberty, teenager, menstrual cycle, adult, parent, menopause, old age and finally death.

Language, words and stories can all be seen as patterns. Compare one language with another and you will be seeking patterns, such as verb formation and grammatical rules. Stories can be collated into patterns, such as the type of story: fantasy, adventure, sci-fi etc., or into the story form such as *boy meets girl*. Rhythm and music are patterns. Having learnt how to play the bass guitar recently, I now understand a bit more about the sections of a song, such as intro, verse, chorus, bridge and outro. Most bass playing involves the repetition of riffs, and it is easier to learn once you see that pattern.

Weather and climate are full of patterns. Every bit of the weather is unique but how could we talk about it unless we simplified it into the broad pattern of what we experience. Maths, numbers, geometry and statistics are bulging with patterns, such as the Fibonacci sequence and the golden ratio. I won't go into details, just go look them up if you are unfamiliar and lose yourself for an evening in amazing YouTube videos and websites, or better still get Priya Hemenway's *The Secret Code: The mysterious formula that rules art, nature and science.*[12] I can't say anymore otherwise I would just be demonstrating how easy it is to get obsessed by patterns. Instead let's keep on track and continue working out why this principle is useful for us and just simply acknowledge that there really are patterns everywhere.

Why do patterns exist?

Well for a start we all live in the same universe that is made up of the same materials and elements. We are all governed by the same natural laws that we have tried to work out over the centuries such as ones about gravity, energy, time and thermodynamics. We all live on the same planet and so we all experience the same natural forces, such as the downward pulling effects of gravity. Then there is the sun and the moon, with their influence on air and water movement, and patterns of daylight, night time and seasons. Time, as far as we can tell, only goes in one direction, and this influences patterns of growth, change and decay. We are all human and we all have something in common, including similar perceptions of the world as determined by our biology and natural selection. The common features of being human – including language and music, and key moments in all of our lives such as birth, change and death – influence many patterns in our lives. Yes there is a lot of ***diversity*** amongst us, including varying perceptions of time, differing ideas about death and many languages, and yet we still have very similar ***observations*** of the patterns around us.

Our obsession with patterns and why we need them

We are obsessed with patterns. Maybe even addicted to them. As humans we see patterns when there aren't any and we add meaning where there isn't any, this is called apophenia.[13] Try a relaxing afternoon of watching passing clouds and you will quickly see some that look like faces or animals. One of my favourites is staring into the embers of a fire. Seeing patterns where there aren't any can be a trap. The theory is that our brains cannot cope with the perception that the universe is incredibly complex. We are hard-wired for pattern recognition because it supports our need for simplicity, understanding and survival. Patterns also support our need for stories, as explored in an article by permaculture designer Liz Postlethwaite.[14] Amazingly there actually is some order in what could be a totally chaotic universe but we need to exercise self-awareness, keen observational skills and discernment when it comes to random pattern recognition. There are underlying patterns in the cosmic order that are inherently life-enhancing, which permaculture design can seek to emulate. We do have a tendency to think in patterns and we need to favour the use of life-enhancing patterns, such as those found in nature.

Our obsession with patterns has led to uniformity, consistency and the repetition of so many towns, houses, high streets and shops. There is an ***efficiency*** of cost savings, ease of control and familiarity with such construction strategies but also over simplification, boredom and at its worst, a repression of creativity and loss of ***diversity***. At the same time there is a pattern to settlements because the needs are common. We all need access to water, resources, food, fuel, shelter, community and protection, and when we get the optimum location of these resources then we thrive. We need to recognise and discern when the pattern is useful and in particular see where the ethics, such as ***people care***, can support the best choice too.

We need patterns. When we see a successful design, we may want to copy it. What we actually need to copy is the pattern of the design. To replicate the details and then place that in a new location with different parameters will set us up for failure. What works in one place may not work in another because the conditions are different. Techniques developed can be very site specific. Put a swale in one location and it works, put another swale in a completely different location and it might look similar but the results may be very different. Patterns can provide inspiration for what might be possible in another location and are better to copy than the details.

Principles are patterns

As I have been writing this book I have been coming to the conclusion that the permaculture principles are patterns. One of the reasons to develop principles in the first place is as a communication tool. To communicate about permaculture's success is to explore why a strategy and technique was adopted in the first place. And from there we find that the principles are the transferable patterns of our best practice. ***Observe*** a site, understand the locality so that you can ***work with nature***, see what ***resources*** are available that you can ***catch and store***, develop your ideas ***slowly***, add more ***elements*** and ***integrate*** them together, bring in ***diversity***, increase your ***yields*** as you can and keep listening to your systems ***feedback***. All of these principles mentioned are requests at the pattern level. They contain no details. They are open to interpretation and context specific.

Learning a master pattern is very much like learning a principle.

Bill Mollison[15]

Patterns as a way to grasp the complexity of a system

Donella Meadows defines a system as 'an interconnected set of elements that is coherently organised in a way that achieves something'.[16] I have some caution about using the word *system* as it can sound mechanistic but it includes nature. When we refer to systems we are acknowledging complexity and how everything is connected both within that system and to neighbouring systems. We can simplify our understanding of those nested systems through our ***observation*** of patterns. When we design we may be focusing on just changing a detail within a system.

For example, we may wish to improve our sleep. The interconnected systems at play could include our patterns of diet, time management, our bedroom layout and seasonal patterns. We will need to observe and consider the impact each of these has on our sleep and see which detail we can and want to change. As psychologist Kurt Lewin apparently once said, "If you truly want to understand something, try to change it."[17] Even once we have made a change we will need to continue ***observing*** to see if what we changed has the desired outcome on the system. This change may manifest as a new pattern emerging, such as finding ourselves tired at the same time every night and suddenly sleeping a lot better. And of course all systems are constantly influencing each other so we need to check every now and again through a feedback mechanism that our decision is still valid.

Pattern languages

Permaculture has been influenced by pattern language writers, in particular books written by Christopher Alexander and other members of his team at the Center for Environmental Structure at Berkeley University. The book this team is most famous for is *A Pattern Language: Towns, Buildings, Construction* (1977).[18] I have found this book to be remarkable every time I have a building project. For example, when I was developing a plan for some new stairs, the book tweaked my draft design so that the stairs became wider, less steep and gained additional functions.

> *No pattern is an isolated entity. Each pattern can exist in the world, only to the extent that it is supported by other patterns: the larger patterns within which it is embedded, the patterns of the same size that surround it, and the smaller patterns which are embedded in it.*
>
> **Christopher Alexander**[19]

Let's look at the example of a house. We can all recall those fantastic childhood drawings of homes, they always include walls, a roof, windows, doors, pathways and maybe a garden as well. These physical aspects alongside other parts, such as heating, furniture and various types of rooms can all be seen as patterns. They aren't the same in every house but they typically are all required to make the house. You will have your own particular details for your house based on where it is, what it is made from, how it is laid out, the number of rooms and what it contains. The house itself is also a pattern, and even broader than that we can say that homes are a pattern. A house is only one type of home, as others might live in other structures: huts, caravans, tents and boats. You can now get a sense that patterns occur at multiple layers, nested within each other and thus so do the details. The point at which we are shifting from detail to pattern or from pattern to detail depends on our perspective of where we are in those nested layers.

Permaculture designers have used pattern languages in all sorts of settings and activities to help map and communicate best design practice. These have manifested as checklists, card decks, websites and books. One of the earliest such pattern languages, although not normally referred to as such, was P.A. Yeomans' Scale of Permanence from the 1950s. This has been through many iterations, with tweaks by Bill Mollison, Dave Jacke and most recently by Luiza Oliveira through the Permaculture Women's Guild, who added in parallel social layers.[20]

Other examples include Dave Jacke and Eric Toensmeier's forest garden pattern language in their two-volume book set *Edible Forest Gardens,*[21] Adam Brock's *Change Here Now: Permaculture Solutions for Personal and Community Transformation.*[22] We also have Rob Hopkins' exploration of Transition Towns as a pattern language from 2010,[23] which in turn inspired pattern language work by permaculture designer, Tom Henfrey.[24]

Tom's pattern language work was primarily for the Transition Research Network, to highlight best practice by researchers supporting the Transition Town movement. The process of devising the patterns was done through a permaculture process over a number of years and resulted in dozens of patterns, such as *building resilience, making a difference* and *space for the unexpected.* They sound like principles, and that is just my point: the principles are not only patterns but a pattern language.

My other favourite pattern language is the Group Works Deck by the Group Pattern Language Project.[25] As a facilitator and previously as a community development worker I love these cards. Each one includes a key phrase, a description, a nice photo, links to related patterns and on the website further details, such as instructions, variations, cautions and even anti-patterns where the opposite occurs. One such pattern is *breaking bread together,* emphasising the importance of eating together to build community and trust.[26] Various ways of implementing this pattern are shared, with the caution to include all dietary types for inclusion and avoiding formal discussion or presentation whilst eating. I like the idea that patterns in this form could be wisdom that we can easily pass down the generations, as a form of storytelling to add to oral traditions alongside song, dance and ritual.

Patterned and rhythmic knowledge is unforgettable.

Bill Mollison and Reny Mia Slay[27]

From detail to pattern

Should we always work ***from patterns to details***? Are there times when we need to start with details and then see what patterns emerge? Permaculture activists love getting stuck into the details of practical action and then move towards sharing the broader patterns of learning. Sometimes we may need to jump in somewhere between pattern level and detail level, other times it may be appropriate to move through the levels intuitively jumping back and forth. We may choose to focus on a particular level of change in a system that will generate the ***greatest change for least effort.***

Emergency situations typically demand immediate attention to details whether that be attending to a cut, a fire or a burst pipe. Get straight to the main focus of attention, deal with it and then maybe later come back to explore the bigger picture and the reasons why the emergency occurred in the system in the first place with the hope of preventing such an incident again.

Statistics are a great way to ***work from details towards patterns***. We may be presented with a vast data set and have no idea, apart from anecdotal stories, that a pattern could be hidden in there. This is what happened when statistician David Spiegelhalter plotted data of sex ratio birth statistics from 1838 to 2012. The resulting graph, which he described as delicious, showed evidence that male births rise towards the end of major wars or economic crises.[28] Another example is when James Lovelock came up with Gaia Theory[29] by looking at massive datasets on global weather conditions and realised they contained patterns that imply homeostasis in the biosphere. More recently a famous graph emerged when climate scientist Michael Mann gathered temperature data sets from ice core samples and coral reef core samples. That resulted in the classic hockey stick graph to show once and for all that the temperature of the planet is changing dramatically, contributing to the evidence that climate change is caused by human activity.[30]

When we are implementing a new project we may choose to ***start small and slow***, especially if the big pattern-level picture is overwhelming. Maybe in our new allotment we just tackle a small corner to begin with and build from there. Soon the bigger vision will emerge of what we want to do with the whole site. I have had a similar feeling with this whole book, and am reminded once again of Myron's *start anywhere, follow it everywhere* principle. Sometimes it doesn't matter where you start as long as you are willing to see where it takes you. This is contrary to the typical strategic approach of starting with a pattern-level vision and a set of goals, then developing an action plan, before going onto the finer details of implementing. Which approach you choose, from pattern to detail or from detail to pattern or somewhere in between, will depend on the context, your capacity, motivations and resources.

The Problem is the Solution

Written in collaboration with Emma Leaf-Grimshaw and Carla Moss

Is that really a problem?

The principle of the ***problem is the solution*** was first mentioned in *A Designer's Manual*.[1] Bill Mollison also referred to it as ***everything works both ways*** in his book with Reny Mia Slay.[2] 'It is only how we see things that makes them advantageous or not,'[3] he wrote. The principle has been included in many permaculture books since with various tweaks on the wording. In her first book, *The Earth User's Guide to Permaculture* (1993),[4] Rosemary Morrow referred to this principle as ***see solutions not problems***, and in his *Permaculture, A Beginners Guide* (2001),[5] Graham Burnett referred to it as ***turn problems into solutions***. We also have Colleen Stevenson's version: ***the solution is within the problem***, which I particularly like.[6] The principle is also sometimes referred to as ***turning liabilities into assets***.[7] Holmgren absorbed this principle along with other attitudinal principles into his principle of ***observe and interact***.[8] Personally I like to keep it, partially for the beautiful way in which it is worded and partially because some great design solutions have been identified by focusing on the effect this principle has on our perspective.

When faced with a situation that could be negatively perceived we are invited to look for the positives and the solutions even if the answer is do nothing. A windy site could be the opportunity for a wind generator, an abundance of slugs could be the opportunity to introduce ducks. This principle is all about a switch in our attitude to look at the positive opportunities rather than being caught up in the negativity of a problem.

Rather than seeing the laborious task of weeding as a chore we can flip it around as harvesting compostable materials for soil fertility. Rather than moaning about the shady garden, we can seek shade tolerant plants that will thrive. Rather than complain about the exposed windy landscape, we can plant a shelterbelt full of edible species. Rather than seeing the boggy damp land as badly drained or the lowland area as a flood risk, instead create a pond or wetland habitat. This change of attitude steps us towards the potential of ***multiple functions*** and increased ***yield***.

We are not the only living beings to have these potential flips in perspective, to see the opportunities in a crisis and solve problems. It is a basis of evolution and survival too, demonstrating aspects of ***adaptation*** and ***cooperation***. Some animals, in particular a range of insects, have evolved to turn poisons consumed when eating plants into defence chemicals or camouflage mechanisms. For example, monarch butterflies store cardiac glycosides from milkweed plants that they eat, resulting in them tasting disgusting to birds.[9] This is a co-evolution as the plants evolve to cope with overgrazing by producing toxins in the first place.[10] Whilst others, in particular various animals, have more direct problem solving techniques, such as the development of simple tools by various birds and apes either through trial and error or by learning from peers.[11]

Problems can become a livelihood. For example, the overgrowth of water hyacinth, which is native to the Amazon and is now clogging rivers in over 50 countries. Ornamental plant lovers have spread this plant globally which has now become an invasive species for many locations. These new locations do not have the predators in place to keep the plant in check, there are no balancing feedback mechanisms. It turns out that the water hyacinth plant provides excellent fibres for weaving.[12] In Cambodia, there is now a company called Rokhak[13] that employs local women to harvest the plant and turn its fibres into bags, rugs and baskets. This creates a win-win situation: supporting the ongoing traditional river livelihoods of fishing by keeping the rivers clearer of water hyacinth. In Bangladesh, large floating garden mats are made from the invasive plant on which crops can be farmed. In Kenya, the plants are mechanically harvested and then anaerobically digested and turned into biogas for cooking. There are many more examples of turning this problem plant into a livelihood solution. It is not all amazing news though, as this plant is a serious costly problem still for many countries due to its rapid growth.

Other invasive species opportunities have included grey squirrels becoming sausages in the UK and Himalayan balsam providing edible seeds. We can view these opportunities as potentially new ***renewable resources*** but that would be a mistake. The risk with these businesses is that they could end up creating a demand for the invasive species and a desire to keep it thriving. There have been reports of water hyacinth cultivation by some basket makers.[14] At the other extreme, the businesses could become so successful that they eliminate the plant completely. In the case of water hyacinth, elimination is very unlikely at the moment.

One ***renewable resource*** that has found a new novel use is hair. It turns out that the natural oils in hair are really useful for attracting and holding onto other oils. A whole new creative environmental initiative called Matter of Trust, has been founded from this insight by using donated hair clippings and turning them into mats, also known as booms, to deal with petroleum spills and other oil waste.[15]

We need to discuss and share our perspectives of a perceived problem before acting. Is it really a problem? Can we identify a solution by tweaking our perspective and attitude? Will our solution make the problem go away or make the situation worse? Can we see a solution where others just see problems? Back in 2009, Abundant Earth saw the opportunity to buy oak logs for building material from a neighbouring forest. The forest owner had assumed that the logs were too wiggly for use as building timber and so they were earmarked for the cardboard industry. We disagreed and hired a sawmill, which generated enough timber for several buildings over many years.

The power of positive thinking

There is a story about a poor donkey that fell down a well. The farmer, shocked and surprised to begin with, realises that it will be very difficult to rescue the donkey. The donkey appears to not be moving and so is assumed dead. The well ran dry many years ago and is no longer in use as the groundwater levels have dropped and he now has a different supply of water. It is a long way down the well and it would be very impractical to lift the donkey out. The farmer and his friends decide to fill in the well with spare soil and bury the donkey. To their surprise though as they start to fill in the well they hear braying and realise the donkey is still alive. The donkey has been shaking the soil off its back and that gives them an idea. They continue to fill in the well and the donkey continues to knock the soil off its back. Gradually the donkey finds itself close enough to the surface until at last it can jump out and trot back to its field.

OK, it's just a story and maybe not a great one. Don't question the details too much! The point is that we can choose how we respond to any given situation. We can see it as doom and gloom, we can see the glass half empty or we can be positive and get creative and find a solution. We can employ the tactics of a growth mindset whereby we are not limited by our skills or experience but know that through effort and work that we can improve and overcome the problem.[16] Having a more positive attitude can also help with our mental health, increase our immunity, lower stress levels and increase our ability to cope and be more ***resilient***.[17] A very simple way to shift towards a more positive attitude is to have a daily gratitude practice. At the end of the day just think of three things that you are grateful for today. Maybe it was a conversation, something you ate or even the roof over your head.

The power of critical thinking

Now we need to balance this positive thinking attitude with some effective critical thinking too. If you are only positive all of the time then you will have an unrealistic perception of the world. We need to pay attention to the ***feedback*** mechanisms around us that draw our attention to a problem in the first place. Being too positive can make us blind to a problem even existing. We need to be able to ***observe*** what the problems are so that we can then identify relevant solutions. It's all very well *keeping calm and carrying on* but sometimes we need to panic and freak out, take a deep breath and find that solution quickly. Barbara Ehrenreich, journalist and political activist, interviewed Wall Street bankers about the financial crash of 2007-8.

In that study she found that those that gave warnings about a pending crash were simply fired by their over-optimistic bosses.[18]

We need to take seriously an individual's perception that they are having a bad time rather than simply telling them to be positive about their troubled situation. It doesn't work to simply tell them to pull up their socks and change their attitude. Someone who has lost their job, experienced pain or grief, needs empathy first and then support to find solutions that work for them, rather than just being told to buck up and get on with it.

We are hardwired to look out for the negativity. That instinctive vigilance and caution to keep an eye out for predators and dangers keep us alive and get us out of the way of danger when the concern becomes reality. There is a balance to be struck and we need to see both positive and negative outlooks and have a healthy scepticism about both extremes.

Critical thinking[19] is a skill that can be developed. It can include the ability to evaluate and analyse information to form a balanced judgement. It is being open minded, curious and yet sceptical. It leads to being able to consider different perspectives and not having a fixed opinion without good evidence and research. We need to understand our biases and be able to put them to one side to identify what is really going on.

Through this skill we can dive deep into a problem to identify an underlying cause rather than just accepting it on a surface level. A perceived problem isn't always the actual problem or issue. Let's consider drug addiction where the drug is perceived as the problem. From a deeper perspective the problem might really be trauma or abuse and the drug is a way of coping with the problem. Therapy might be part of the solution. A really useful tool to get to the root cause of a problem is to simply ask why and then repeatedly ask why to each consecutive answer until you get to the root cause answer. This is sometimes called the 'five whys tool', although it doesn't need to stop at asking why for the fifth time – maybe the sixth or even seventh time might generate an even better response. It was developed by the Japanese inventor, Sakichi Toyoda, and is part of the lean philosophy used by Toyota.[20]

Many people new to permaculture will start by growing some vegetables. It can be tricky, and we may discover that we don't have 'green fingers'. Turn on the growth mindset and be patient and develop your skills. At the same time know that improving soil fertility, finding more suitable plants, choosing a different planting plan or a different seed supply could all help. The problem isn't a negative thing, it's an opportunity to look more closely at the challenge.

Tools that can help support and map our critical thinking could include Edward de Bono's tool PMI[21] or the SWOT tool,[22] developed in the 1960s. PMI stands for plus, minus and interesting. SWOT stands for strengths, weaknesses, opportunities and threats. There is a common variation of this tool used by permaculture designers called SWOC, where the C stands for constraints or challenges. Notice that both begin with a positive header. This is no accident. Edward de Bono made it clear when he developed this tool that as humans we have a tendency towards the negative first. He deliberately put the plus first to encourage us to start with the positives as we can easily jump to criticising straight away instead. The purpose of the PMI tool is to avoid jumping to any conclusion but instead give time to consider a situation from a range of perspectives both positive and negative and anything in between. You can see an example of PMI in action in the ***cooperation*** chapter where I compare cooperation with competition.

Positive Internal Factors Strengths	Negative Internal Factors Weaknesses
Positive External Factors Opportunities	Negative External Factors Threats

SWOT analysis tool template

Finding the unexpected resource

Back in 2001, before we had bought our smallholding, I was mainly doing private and community landscape projects. One of those projects in Benwell, in the west end of Newcastle, involved developing an abandoned plot to the rear of a community centre. The project had limited funds but had already had a landscape architect's design done. This design failed to notice a key feature. One of the first steps for any permaculture designer is to ***observe*** your site and that can involve doing some sample digging. Clunk went the spade repeatedly just a few inches below the thin dusty soil. At first I thought this was an entire layer of solid concrete, which would be a problem for planting any trees or shrubs. It turned out to be grasscrete, concrete blocks that contain spaces to allow the grass to grow through.[23] Thus the surface can end up looking green and can enable the space to be used for parking. The laying of these blocks predated the project by many years and they had no idea of their existence. The project had asked for raised beds to be constructed and I had doubts that they could afford the materials on the limited budget. The problem had become a solution. The grasscrete blocks could be lifted and each one was the size of several standard bricks. They could be used as the building material for the raised beds.

Raised beds made with grasscrete and planting in progress © Wilf Richards

There were hundreds of them, allowing me to create several raised beds to a reasonable height. There was also sand, broken bricks and rubble available on site that could be placed at the bottom of these raised beds to enable good drainage. Thus no skips were used to take the rubbish away, ticking the ***produce no waste*** principle at the same time. After top soil and plants were added, the garden was complete, and the build was on budget too.

We can find unexpected resources in the 'problem people' that we encounter too. You will have encountered that person that really annoys you. You may think that your choices are limited to reacting badly or avoiding such a person. But apply the ***people care*** ethic and give that person some time, listen carefully, get curious and maybe provide some empathy. Maybe they are just having a bad day. Maybe there is a trauma there or a neurodiversity that limits their social skills. Maybe it turns out that they have a skill you can admire and praise. Maybe they have something to teach you. Maybe you have something in common. They might even turn out to become a great friend.

Produce No Waste

Written in collaboration with Cathrine Dolleris

There is no such thing as waste in nature

We have been steadily developing a waste problem with major effects on our environment, so it is no surprise that permaculture designers, along with nearly everyone else, really want to find solutions to the waste problem and also contribute to the permaculture ethic of ***earth care*** at the same time. Bill Mollison had an interesting angle on the notion of waste:

> *Any system or organism can accept only that quantity of a resource which can be used productively. Any resource input beyond that point throws the system or organism into disorder. Oversupply of a resource is a form of chronic pollution.*
>
> **Bill Mollison, Principle of Disorder**[1]

Bill was mainly considering resources here that would at face value actually be totally useful. For example, an oversupply of food would be waste and thus a pollutant. We can also expand that notion to carbon dioxide, at the right level it can be reused at a steady rate but get it out of balance by pumping lots into the atmosphere too quickly and we end up with global warming. He wrote about careful management of resources to minimise waste and restrict pollution as part of our responsibilities.

After all, we don't see waste in nature so why are we wasting so much. Or do we? There are some examples of waste in nature. Cats half-eating mice, plants generating too much sugar and exuding it onto their leaves and the recent spate of orcas found to be only eating the livers from certain sharks. There are large-scale cases too, such as the release of oxygen from phytoplankton millions of years ago. This excess oxygen was a waste to them but in turn it changed the atmosphere on this planet to enable the evolution of animals, so not wasted after all. In nature there is always something else that will come along to finish off what's been left behind. Even our modern waste will rot and transform. It just might take a very long time and be terribly toxic to other life in the meanwhile. Not great!

Any given organism might have too much of something at times, or need to dispose of that thing it no longer needs, or the substance may be toxic to it. The obvious safer examples are animal wastes, such as faeces, or the shedding of leaves from a tree. In such cases the waste matter deposited is dealt with by another species relatively quickly or by the actions of sun, wind and rain. Bacteria, fungi, nematodes and many insects, bugs and animals need these wastes as part of their core diet. This is part of our understanding of ***working with nature***, or as Graham Burnett put it[2] ***everything cycles.*** Permaculture designers are keen to establish systems, such as compost toilets, whereby any wastes we do generate have the potential to be energy for something else. This is why in *Introduction to Permaculture*[3] by Mollison and Slay, there were a couple of principles introduced to make this clear.

Those principles were ***using biological resources*** and the ***cycling of energy, nutrients and resources.*** If we are using biological resources in the first place, as compared to industrial or processed chemical resources, then all of our wastes should be biodegradable, recyclable or easily become food for the next in the chain. And the focus on cycling of energy, nutrients and resources is largely only possible if that form of energy is biological in the first place. Examples given in the book included turning kitchen waste into compost, biogas production from animal manure and utilising household grey water in the garden as fertiliser. All far better than using non-renewable fossil fuel based resources which might be toxic to soil, water, air and life in general.

In Holmgren's set of principles he added extra clarity to this by coining the wording ***produce no waste.***[4] He asks us to value and make use of all resources so that nothing goes to waste. The applications clearly broaden and shift here towards recycling, reusing and maintaining what we already have, but clearly also still have a strong focus on using natural biological resources and cycling nutrients. Holmgren regarded ***efficient energy planning*** as one of the origins of produce no waste because by careful placement of elements we can enable outputs to become inputs and eliminate waste. Other examples of this principle have included ***prohibit waste*** (Heather Jo Flores),[5] ***waste nothing, recycle everything*** (Sarah Spencer)[6] and ***waste nothing, inputs and outputs*** (Neil Kingsnorth).[7]

What a load of old rubbish

In Brighton where I grew up in the 1980s, there were plenty of derelict buildings to explore and turn into squats. One encounter I will never forget was finding a printed sign in one of those abandoned buildings amongst a pile of rubbish. I can only assume it had been left by an art student or maybe an archaeologist as a joke. It simply said, 'You can tell a lot about a society by what you find in the rubbish.' I wish I had a photo of it, it was an act of genius as far as I was concerned at the time.

In researching for this book I presumed the problem of waste began in the industrial growth era but I was wrong, as any archaeologist will tell you that dumping waste began in the prehistoric era. Landfills have been around since at least 3000 BC, and simply consisted of pits covered in earth.[8] In Britain in the 1300s, the first rubbish legislation was introduced. The issues as a result of these early rubbish systems were largely minimal as populations were smaller, the amount of waste generated would have been less and the likelihood of the waste being toxic was low and almost everything was biodegradable. The main problem would have been the control of rats and polluting waterways with manure and poo. As populations and cities grew in number and size, these issues became unbearable and so in the 1800s many European countries introduced health-related laws to manage street collection of waste, development of large-scale landfill sites and development of the sewage system. This did not solve the problem though, it just shifted it along and out of sight.

Jump forward several decades into the 20th century and we have taken the strategy of landfills to another level. The sheer quantity of waste from a growing population resulted in the filling in of valleys and opencast mines, resulting in toxic water, run off that kills life and methane generation that has a faster impact on climate change. In the 1970s the responsibility for dealing with waste was moved away from the industries that produced packaging to households and consumers. This led to the recycling drive in households and a burden upon local authorities to collect and deal with single-use items. With the rubbish being taken out of sight, littering became an issue, which has been shown to be contagious and cultural. The more litter you see on the streets the more likely you are to litter. It is also easy for plastic items to simply blow away. We have so many types of single use plastics now and they are very hard to recycle. Massive quantities of plastic end up in the Pacific Ocean in an area twice the size of Texas. The scale of the problem of waste is enormous and so varied.

One man's trash is another man's treasure

Back in the day, some of the waste issues were lessened by hoarding, scavenging and storing away for a rainy day. Being a scavenger has been a serious profession in the past but has since been made illegal in many countries. Scavengers have long demonstrated the ***problem is the solution*** by creating livelihoods from another person's waste. The benefits of scavengers to tackle the waste issue is recognised positively by some countries.[9] After all, *waste not, want not.*

Throwing things away is an attitude that has developed over the last century. Plenty of us will have parents that have a garage and even an entire house full of bits. My friend Phil has always referred to these stored bits as scratch, so that he can *start from scratch* when he wants to make, fix or create something. There are plenty of opportunities for artists in using waste.[10] When an item comes to the end of one life, parts might be identified as potentially reusable. You might have limited space to store your scratch, maybe only a cupboard or a garage. That ***limiting factor*** is just as well because if you have a smallholding then the storage of bits can get out of hand. We love storing wood, metal, nails, screws, glass, car parts and lots more. If we are going to use what we have already then we need to make sure we have a good stock of things to use. Every now and again we have to have a clear out when we realise those components can't or won't be reused and it's time they were passed on to the recycling centre. There have been plenty of moments though when those bits find their new home and it's a joy and a sense of ***efficiency*** to know I have spent no money and reused a part that was waiting patiently for its next life. The prospect of clearing out the parents' garage after they pass away though comes with mixed feelings. There might be treasures but there might also be thirty-year-old electricity bills (I kid you not!). There is a balance to be had in the desire to hoard.

The right to repair has been gaining strength in recent years as a backlash to the corporate desire for absolute control of its products. This is merely a resurrection of an age-old approach. Things were built to last and they were fixable. *A stitch in time saves nine.* The growth of repair cafes and repair events in the last decade is welcome but these services used to be a trade. The last household goods repair shop in our nearest neighbourhood closed in the Covid years, although I am pleased to see that many sewing and stitching based livelihoods appear to still be thriving.

If an item was no longer wanted but still in good use, they were taken to the jumble sale or the charity shop. Jumble sales seem a thing of the recent past even to me and charity shops struggle to find quality items as they will be sold through an online platform instead. Cultures around the issue of waste have shifted in the last generation but the issue of waste has not gone away.

This is not just your problem!

We can all make significant steps to reduce our waste personally and we could reach the no-waste nirvana in our own household but I see a greater power in this principle when we apply it at scale across society. We can demonstrate this with a bit of maths.[11]

For example, there is a far greater overall reduction in waste if 50% of the population reduce their waste by 10% each, as compared to 1% of the population reducing it by 80%. Let's take a sample population of 100 people and say each person produces 1kg of waste each week (it's obviously lots more than that but this is just a maths exercise). So that's 100kg of total waste generated. If 50% reduce their waste by 10% then 50% are producing 900g each and 50% are still producing 1kg each. Total waste generated is now 95kg.

Alternatively if 1% reduce their waste by a massive 80% then they have reduced their waste down to just 200g. Amazing, but the other 99% of people are still creating 99kg a week. An excellent effort by one individual but you can see from the numbers that we need to utilise society-wide strategies to create an ***efficient*** response to match the scale of the problem. So practise ***people care*** by not beating yourself up when you don't reach that zero-waste nirvana. We need to get on board society-level strategies instead.

The Rs of efficiency

At a society level we are encouraged to follow all of the Rs in a particular order as follows: reduce, reuse (including repair and repurpose), recycle and recover. Only after all of those options have been tried should we consider disposal of waste. This is known as the waste hierarchy, of which there are various versions. We are persuaded to follow these steps so that landfill becomes the last option.[12]

We can add another R to the list: redesign. Many products need redesigning in the first place so they can be repaired, reused or recycled with ease. This would be a step towards the cradle to cradle approach such as the Fairphone[13] where the whole phone is designed to be reused or recycled with components easy to replace by the user. When you need to replace an old or broken item, spend some time researching the best option to shift towards no waste. Systems are being redesigned too, some tech and household item businesses are shifting the nature of their business from selling the product to leasing it instead. In that way the company still owns the item and will receive back faulty goods that are easier for the manufacturer to recycle, repair and reuse.

If it can't be reduced, reused, repaired, rebuilt, refurbished, refinished, resold, recycled or composted, then it should be restricted, redesigned or removed from production.

Pete Seeger[14]

To make appropriate changes we need incentives from business and governments such as taxes, laws and subsidised schemes. Households in Belgium now have microchips on their bins that enable a precise financial charge to each household based on amounts recycled and amounts going to landfill.[15] Even when we do follow the government schemes, we are very detached from the processes and trust is thin. Governments and businesses don't always understand where their own recycling systems end up. In a devious bit of research, featured on the Bloomberg YouTube channel, electronic GPS tags were attached to three different plastic items that were then sent for recycling via a Tesco supermarket.[16] These items travelled for miles ending up in some unusual places, including becoming fuel for a cement factory in Poland. So not recycled at all, although the use of plastic waste for fuel is driving down the use of coal.

The task of recycling on these large scales is not easy; it is fraught with complexities, lack of traceability and often results in the pollution simply being shifted to somewhere else. The incinerators that burn the waste and generate heat and electricity have been growing in numbers but again the pollution and toxic residues are very undesirable. Some recycling only makes sense when it is highly centralised, such as gold extraction from electronic products, which is energy intensive and involves a lot of transportation. Meanwhile we now have proposals for mining landfill sites to extract materials buried from years ago. What a mess this all is. So what can we do as humble permaculture designers?

Inputs and outputs

As permaculture designers we love exploring and identifying waste in our systems. We see these potential pollutants as unused resources. They are inputs to another system or even back into the same system. We just need to work out where that resource can go safely and be of maximum benefit. We can use the input-output analysis tool to help us see how outputs from one system can become inputs to another system.

Being aware of your waste products and your neighbours' waste maximises opportunities to turn those outputs into inputs. Always try to find a new purpose for the unwanted item before disposal. And always seek the secondhand option before buying brand new. When looking for something to weigh down our weed barrier membrane in our market garden, we chose old car tyres as they were free, could be delivered to us and were a waste problem to the local mechanics. Although they are not the most attractive of items they do an amazing job of keeping the membrane in place during strong windy days.

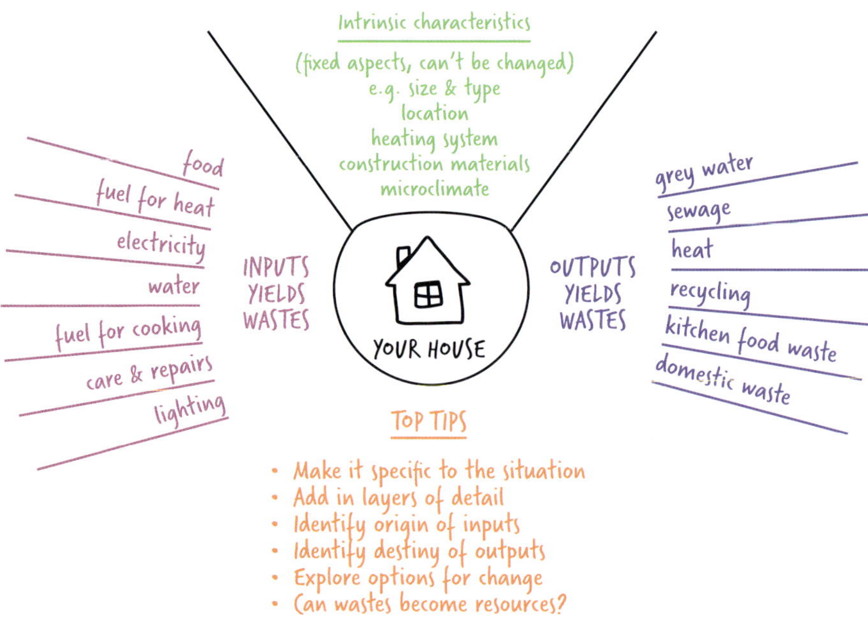

Example input-output analysis for a house © Wilf Richards

We can make significant steps as an individual to reduce our waste or pollution. For a start, anything that lowers your carbon footprint is a win win. At a home level it could include setting up a compost system to reduce your waste. Typically about half of our waste can be composted. We can choose items that come in compostable packaging, or use refill shops or buy wholesale to minimise packaging. We recently switched to a different coffee provider because their packaging can be put into our compost system. When it comes to cooking we can consider what needs eating up first, only buying what we need in the first place and being aware of sell by dates to minimise food waste. And if you can, get a wormery or bokashi compost system because they will deal with cooked or out-of-date food waste. When you do need to go shopping or out and about, then remember that reusable bag you love and the water bottle that can be used over and over again.

When you do need to buy brand new then look at your options and consider what will happen to the item when you are finished with it. Consider its whole life cycle. In the building of our home we have ensured everything is biodegradable, compostable or can be reused. This has included straw bales for walls, clay and lime render, insulation made from sheep's wool or newspaper and local timber where possible. Off cuts of timber are stored as *scratch* for potential future use or if small enough and old enough will become firewood. This approach ensures that we are taking responsibility for what could otherwise become waste for a future generation.

There are plenty of opportunities to focus on maintaining and repairing an item before we need to repurpose it, upcycle it, and then and only then, consider recycling. On our smallholding we have various ***integrated*** aspects that enable the waste from one system to become a resource and input to another area. For example, the sawdust from the sawmill is used in the compost toilet. The humanure from the toilet is composted further and then used as a fertiliser for fruit trees and wood chips from the planing machine can be used as bedding in the hen area or as mulch for pathways. Nothing is perfect though and we still struggle with some plastic based items, such as nets, punnets and storage bags used for wholesale veg. We have tried switching to compostable bags for packing salad but the quality is poor, they are expensive and not designed for reuse.

In our communities we can head to the local waste food cafe, use the local repair shop or the community composting scheme. And if you don't have any of those things in your area then that could be an opportunity to establish one. There are also opportunities for creativity with our waste as a source of art material, and to increase our creativity we need to do some relaxing and waste some time! After all, the ***yield is only limited by your imagination.***

Relative Location

Written in collaboration with Neil Kingsnorth

Who coined this one?

As far as I can tell the first mention of ***relative location*** is in the article by John Quinney in 1984.[1] He clearly describes it as a principle that is applied as part of the consideration of zones and sectors in a site design to enable the saving of labour and energy. Bill Mollison and David Holmgren shared details about zones and sectors in the first permaculture book back in 1978 but the term *relative location* did not appear in any books until *Introduction to Permaculture* in 1991.[2] The term was also used by Sego Jackson in winter 1984 in an article he wrote for the *Living with the Land* magazine.[3] Maybe Bill had used the term in conversation or in teaching before 1984 but as it stands at the moment I think we may need to give credit to John Quinney for this principle.

Distance invites neglect, while proximity encourages management.

John Quinney

I want to clarify a potential confusion here as this chapter explores the design tool known as *zones and sectors*, and this tool typically features in permaculture books around the principle of ***efficient energy planning***,[4] which also appeared in the *Introduction to Permaculture* book alongside relative location. The overlaps between these two principles are clear as they are both about placement. Consequently I am including all of those things here in this chapter. Just don't get this confused with ***energy efficiency***, which overlaps but isn't the same.

The term *relative location* brilliantly and simply encapsulates one of the fundamental principles of permaculture. This is the idea that the success and behaviour of any one part of a system is influenced by, or depends on, its relationship to any other part of that system. The distance between two elements, their *relative location*, will influence the type and strength of the connection between those two elements. As with most permaculture principles, the concept comes from an observation of the behaviours of the natural world, and this principle can be seen at play at the large scale, right down to the smallest.

At play in the natural world

Relative location is perhaps most easily observed at the small scale in the natural world of plants. Most plants have particular needs and grow strongest when those needs are well met. Bracken grows vigorously on sunlit, moist land, but cannot grow in deep shade, for instance. Many wildflowers grow well on poor soils but are outcompeted by more vigorous plants on richer, more fertile land. On waterlogged land, creeping buttercups can outcompete even strong grasses. Animals will thrive when they occupy a habitat close to their core needs of water and food, and we can see this in the early development of human habitation too as villages become towns where the resources are conveniently nearby.

The conditions created in any one space on a piece of land are influenced by everything that is around that space, and is arriving into that space. There are many influences upon a space, such as the shade cast to the north of a tall tree, the frost pocket created by a fence or rock, the compaction of the land created by a nearby path or the turbulence created by a large shrub.

Imagine a mature oak at the edge of a woodland. The presence of its trunk might create conditions for ivy to thrive by climbing it. On the north side of the trunk, where it is shadier and wetter, there might be mosses growing that cannot grow on the sunnier southern side of the trunk. Down on the ground, to the north of the oak in summer, its leaves will cast deep shade and ferns may be thriving. On the southern side meanwhile, a much greater diversity of plants may be enjoying the sunlight, and some may be thriving best in the shallow soils around the roots that others can't get a foothold in.

In other words, the success of any one element in a system depends on, is influenced by, and also influences, other elements in that system. The influences exerted outwards by a species are also given consideration in the ***everything gardens*** chapter. Here though, with relative location, we are more concerned with the influences and interactions between the elements in a design setting. This is the basis of relative location. The location in which we place elements in our designs will influence how that element interacts with neighbouring elements. Just as happens in the natural world, we view elements as connected rather than each one isolated. There is an absolute joy in getting to apply this principle as the ***energy*** saving measures can be amazing. There are a variety of ways of applying relative location.

Zones and sectors

The concepts of zones and sectors appear in the earliest books about permaculture, although that is not the first time they appear in literature as the model has its roots in the spatial economic and zonal analysis of rural land use by Prussian economist Johann von Thünen in the early 1800s.[5]

Zoning is a simple but very powerful tool that helps separate a site out into different areas based on how much labour, time or attention that area needs, and thus we can use it for the purposes of ***efficiency*** and when considering the relative placement of elements. Zones are normally focused from one central point, such as your home or yourself, which you might describe as your 'centre'. Elements that need frequent visits or are labour intensive need to be placed closer to that centre, whereas those requiring less input can be placed further away. This is why this tool is normally associated with the principle of ***efficient energy planning***.

For example, annual crops that require regular watering, harvesting and pest control are placed closer to the centre, whereas tree crops that only require annual harvests and annual pruning might be placed further away. Furthest away from the centre of activity will tend to be the wilder areas of a design. At a large scale, that might be a whole nature reserve. In a garden, it may be the wildlife corner.

The zoning tool can also be used in social settings and the application of it will vary enormously according to the context, situation and designer. The most common application is in regard to family and friends, whereby we place immediate family closest as they require daily attention; whether that be your child or partner. Further away, yet hopefully close by, we find our friends and maybe other family members in the neighbouring houses or nearby towns. Many of our colleagues and friends might be close by but typically the further away someone is the less likely we are to be able to maintain that relationship even with modern technology.

The term sector, meanwhile, refers to the energies that cut through zones, arriving, leaving and flowing through a site. These can be physical energies such as sun, wind, fire and water. They can also be subjective or social in context such as a view or noise pollution. Once we have identified and understood the influence of these various energies, we can choose in our designs whether to block, enhance, exploit or just observe these energies. For example, wind could be blocked with a shelterbelt, or exploited with a wind turbine. A view could be protected and framed if it's enjoyed or screened out if you don't like it.

Knowledge of zones and sectors has play a considerable influence on the position of elements that we decide upon in our designs. This use of relative location will help us to place our elements where the mutual benefits will be maximised and ***energy*** will be saved.

Minimal work

Our placement of elements in a system may take advantage of an external energy or element that will do work for us or save us energy. This might be, for example, gravity, a regular passing person or an animal doing the work for us because it is part of its nature. We could regard this type of implementation as a form of ***energy efficiency*** or if the focus is a change or maintenance of an existing system then it may overlap with the principle of ***least change***.

A very early example of this appears in a 1930s book about building privies (outdoor toilets to you and me) called *The Specialist* by Charles Sale.[6] Of course it does not use the term relative location but it certainly refers to the energy saving placement and relationships between people, wood store, house and toilet. In those days (and still for some of us), toilets were placed outdoors and not right next to the house. By deliberately placing the woodstore in between the house and toilet then there would be ease in the regular collection of firewood into the house, as every time someone goes to the toilet they have the opportunity to bring back an armful of firewood.

Systems that use gravity to shift water are also an obvious example here. If you have a rainwater harvesting system, ideally you want it as high in the landscape as possible so that the water can be gravity fed to where it is needed. This has influenced our choice of placement of buildings on our smallholding which has no mains water. Our market garden depends entirely on rainwater systems. Thus the rainwater harvesting systems are placed carefully uphill from the market garden to utilise gravity fed water which means we don't need to install electrical pumps.

The other example commonly referred to in many early permaculture books is the food forest for hens. By placing our hens into a landscape full of perennial fruits, herbs, seeds and nuts we can lower the cost of external feed for the birds. Patrick Whitefield referred to this practice of putting the food and consumers in the same place as ***relative placement*** in his book *Permaculture in a Nutshell.*[7]

We can also save energy when a key resource, such as water, is required by multiple systems within a locality. For example, if our market garden, home, livestock and greenhouses all need a water supply then we might consider where that

Summer house at Abundant Earth with rainwater harvesting © Wilf Richards

water is arriving onto the site and how far away each of those systems are from each other. If it is possible we can minimise infrastructure costs by placing these systems closer to each other when we create the design in the first place.

Outputs become inputs

Relative location can be a way to ease the flow of energy from one element to another. Permaculture designers are interested in the inputs and outputs of each element that we place in a design. We can help the outputs from one element or system to easily become the input to another element or system simply by the way the elements are placed as we saw in the ***produce no waste*** chapter.

For example, you may want to put your compost system near to your back door for ease of access for emptying the kitchen caddy (assuming there are no issues with rats). You are likely to find it much easier to empty your caddy if the compost bay isn't too far from the kitchen, especially on a rainy, winter's day.

Complex relationships

When I redesigned our compost system for our market garden in 2014 I had an opportunity to relocate the compost bays. Since it started, the market garden had expanded sideways and the compost system needed to be more central in the garden. We chose a location next to the track for good access and in a shady place that was proving not great for veg growing. This relocation made it easier to get materials in and get the compost out. Afterwards in this same area we also added a sawdust urinal toilet. The output of the urinal toilet could then be emptied with ease into the compost system. This was primarily influenced by the principles of relative location and ***produce no waste***.

A nice example of relative location at play is in field barns, which are common in Northern England. The barns can be seen in remote locations across the countryside often in the middle of fields. They were placed very carefully in the landscape, often far away from the farmstead, with ***multiple functions*** in mind, because they needed to maximise the beneficial relationships between landscape, meadow, pasture, hay and cattle. The barns would store hay in the loft, house cattle on the lower floor in the winter and act as a central place to gather dung from the cattle whilst housed there. The dung could be removed in the spring and used to fertilise the meadow field immediately surrounding the barn. At this point the cattle would be moved uphill for summer grazing in upland pastures whilst the meadow grass grew. Once the meadow was cut in early autumn and stored in the barn loft, the cattle could return to be housed downstairs in the winter once again. The barns are typically on sloped ground with access to the hayloft from the uphill side and access to the lower floor from the downhill side. Gravity is being used as a friend too as the hay is simply passed down through a hatch to the cattle below.

Field barn in North Yorkshire © Stephen Andrews

When we considered the placement of our market garden in our smallholding the first relationship was clearly with access to water. Because of this, it made sense to start growing vegetables next to the only building on site at the time so that we could ***catch and store*** rainwater. Since we started the garden in early 2002, it has grown in size. It turns out we chose the location well and have expanded on it as our needs have grown and as the benefits of the location have become more and more obvious. We started ***small and slow***, but soon realised that the location had the best microclimate on site as well. It is also near to the front gate so starting a veg box scheme on the same location for ease of deliveries on and off site was straightforward, and convenient for volunteers to access the site too. This incremental design was strengthened by the step by step recognition of relative location. There were times when we considered moving the veg garden in the early days but when we discussed the relationships required to maintain it we realised it was best just where it was already and could be expanded instead. We also used this principle to support the location of our pick-up points for the veg boxes, choosing locations with good access and typically a house where we could install a storage box that could be accessed at any time rather than at a cafe or shop that had limited opening hours.

We can see in these examples of complex relationships the dynamics of natural systems in play, the development of finding the right niche and lots of overlapping ***edges*** being encouraged in the placement of elements in systems.

Keeping your distance

Of course, ***everything works both ways***, in other words sometimes the relative location of elements can be a problem rather than a solution. There are times when we recognise the energy or the output from one element is antagonising a nearby element. In these situations distance is required to minimise negative relationships. In these situations we are still considering relative location but ensuring greater distance between the two elements. This is similar to what we have already explored in the ***integration*** chapter, in other words it is sometimes better to segregate some elements. Interestingly in an early exploration of relative location Holmgren brought it under the heading of his principle focused on integration.[8]

The most common example on our farm here is the herbivore versus the vegetable garden. If it is at all possible then my advice is to keep two fences between where your livestock live and where your vegetables grow. That is not possible at our site and so we have to be super careful not to leave the pasture gate open to the market garden, thus preventing the sheep from gaining access. It is one of the reasons why we don't have goats as their ability to climb a fence is remarkable. Most hen enclosures will need high fences to keep foxes out and most gardens will need rabbit proofing, for many of the same reasons.

We may also see the need to maintain distance in town planning design. Noisy industry is better placed away from quiet residential locations. Late night opening clubs and bars need to be considerate of their residential neighbours or find ways to minimise the leakage of noise at certain times. Equally the placement of schools, community centres and other key services needs to be within easy reach of local residents to maximise their use and develop ***multiple functions*** for those locations.

We love to have volunteers involved in our smallholding, and yet at the same time we need to manage when they are here. On the days of packing veg boxes we need all our brain power to monitor quantities and maintain quality and so we actively discourage volunteers on those days to minimise distractions. Equally we mainly keep weekends for ourselves and leisure activities. Even though those could be good times for many volunteers, it rarely suits our need for a bit of rest and relaxation after a busy week.

Get creative

The purpose of relative location is clearly to maximise beneficial connections and minimise negative connections. To explore potential new connections in a creative way that pushes you out of your normal thought patterns, you can turn to a great design tool called Random Assembly, which first appeared in the *Designer's Manual*.[9]

This tool makes use of prepositions, which are the connecting words we can use to bridge between any two elements in a system. Those words include in, on, under, over, next to, between, beside, attached to, containing and many more. The game is played by using two sets of cards, the first set stating the connecting words, and then a second set of cards that illustrate the items that could be placed into the design i.e. the elements. We then start to generate random connections between two different elements by using the connecting words.

The process helps us to think outside of the box and break normal patterns of behaviour. Many of the options generated might actually sound crazy to begin with but some such as 'pond inside greenhouse' turn out to be super useful. Even ridiculous outcomes from the process (such as 'pond over house' perhaps) can trigger a useful thought or idea. More likely you might want to actually place your greenhouse on the sunny side of your house where the wall of the house will keep the greenhouse warmer overnight and the whole location provides for easy access from the house to water seedlings.

Location, location, location

Permaculture as a process encourages reflection, tweaking and double-checking. Designs can evolve and it's useful to assume that a design is open to change, even after everything is down on paper. Our ***observations*** need to continue beyond the placement of any ***elements.*** Did we get it right? Can we improve the system even more? Can we take advantage of a passing element? Can we utilise gravity?

When it comes to location, our houses and what surrounds them can be a key use of relative location. Put your house in a large open space and it could be vulnerable to wind and extremities of temperature. Place it amongst deciduous trees and you will find cool shade in the summer and still allow the winter sun to reach your house.

Can we improve the beneficial relationships by moving an element? Can we minimise negative connections by moving certain elements further apart, geographically, or over time? Can we find the best location for any given element to minimise energy input and maximise energy output? Maybe ***optimum location*** would be a better name for this principle. That said, systems are complicated things and perfect can be the enemy of good. There comes a moment when it is time to bite the bullet, start to implement your design, and then learn where there may be further benefits from relative location as you go.

Renewable Resources and Services

Biological resources

The early books of permaculture refer a lot to the use of biological resources but it is not expressed as a principle at that stage. These early examples are based on ***working with nature*** and include the use of animals, manure, compost, trees, soil, sun, water and air. The focus is on shifting away from fossil fuels and instead towards natural processes. For example, why use oil-based pesticides when you can encourage wildlife to tackle the bugs for you. This is seen as simply being more natural and living just as nature does, using the living resources that are immediately at hand just as all other life does.

In the *Designer's Manual* book[1] Bill Mollison writes a dedicated section about resources. He explores them as the flow of energies arriving into a system, being used and then leaving either as ***waste*** or generating ***yield***. Our aim as designers is to generate increased yield by being conservative in resource use and minimising waste by turning them into a resource. Bill regards the forest, soils, air, water and sunlight as natural resources and part of our common heritage. He categorised resources into several types including those which increase by modest use (such as coppicing in a woodland), those unaffected by use (such as a view), those which disappear if not used (annual crops for example), those reduced by use (fossil fuels) and finally those which pollute or destroy other resources (e.g. poisons and concrete). From this he developed a policy of resource management.

> *A responsible human society bans the use of resources which permanently reduce yields of sustainable resources e.g pollutants, persistent poisons, radioactive, large areas of concrete and highways, sewers from city to sea.*
>
> **Bill Mollison, Policy of Resource Management**[2]

It is in the 1991 *Introduction to Permaculture*[3] book that we see this principle become ***using biological resources***, and acknowledged as sourced from John Quinney's 1984 article.[4] Both the book and the article focused on using plants and animals to do work for us with the purpose of saving ***energy***. For example, plants and animals can provide fuel, fertiliser, pest control, nutrient recycling and weed control, such as using nitrogen fixing plants and green manures instead of artificial fertilisers made from fossil fuels. At the same time the use of plastic pipes, tractors and earth moving machinery received no objection in the description of this principle as long as they were being used to develop a long term biological system. This was a pragmatic approach as the transition to sustainable systems can be hard. There are also plenty of details shared about using animals as a biological resource with cautions included. For example, timing is key to utilising an animal, such as a hen for pest control in a veg garden. Bring the animal in too soon and it will trample and destroy the young plants, leave it in there too long and it will eat the harvest. Since those early writings on this principle more examples of its application have emerged over the years including horses extracting timber, planting living willow as a fence (called a fedge), planting trees near houses to help with shade instead of using air conditioning and stabilising eroded river banks with willow spiling.[5] All of these are trying to increase the ***yield*** of a system, reduce the human work required, decrease the need for external inputs and reduce pollution.

I am not a number

There is a tone in these early writings that feels in conflict with the ethics to me. I think it is important that we don't see animals and plants as resources. They are living beings needing respect and autonomy. This is a point where it is essential to apply the ethics as well as the principle, especially ***people care*** and ***earth care***. I see plants and animals as people too and we need to care for them and respect their will and desire to live a natural life too. Avoiding exploitation of animals is a step but is it possible to go even further and avoid exploiting all species? We enter into tricky ethical territory here about what is right and wrong, for example, is it ok to use special breeds of working animals for traditional heritage activities, such as shire horses and oxen pulling a plough? On one hand they seem to love the work but on the other hand they do not have any choice in the matter.

Mother hen and her chicks © Wilf Richards

Having been a hen keeper, I struggled to square the deal of protecting and feeding the hens in exchange for their eggs and manure. Having to dispatch them when they were at an end of life moment or when there were too many cockerels never felt right. I no longer keep hens. I still buy organic eggs for our veg box customers but with every year I feel less certain of this as I learn about the climate impact of feed, even organic feed, and know the move towards being even more plant based in our diets is important. One argument in favour of plough horses over tractors is that horses are regenerative, in other words they can breed and replace themselves long term. Although this does also apply to hens, there is the problem that male chicks do not provide eggs, and although they can be kept for meat, the efficiency is marginal, which is why commercially all male chicks are killed within 24 hours of hatching. Not great!

There are many wild species that are suffering decline so if we can provide wildlife habitats as part of our farming mix, and some of those wild species happen to also be friends in our gardens then that's a win win. Careful observation of what wildlife is managing your pests for you and designing more habitats or protection for those species is a possibility. This might include biological control strategies, integrated pest management, companion planting or creating wildlife areas.

The ethics need to also be applied when it comes to our relationships in working with volunteers and staff. It can be all too easy to take volunteers and members of staff for granted, treating them as just a resource to be exploited and used. Gratitude is key with regular celebration and appreciation of everyone's efforts to keep a system going. Those businesses that do not value their staff and show respect have a high turnover of employees, costing the company dearly in recruitment time and reputation.

Valuing renewable resources and services

David Holmgren evolved this principle into ***use and value renewable resources and services*** in his principles book.[6] A renewable resource is one that is not depleted as it is used or it can be regenerated as quickly as it is used and it should also not have a negative impact on the environment. This is a core definition of the word sustainable. For example, if we build a house from timber, then in theory we can grow the trees required to replace the house before that replacement is actually necessary. Compare that to steel and concrete from mined resources that took millions of years to form. The more we use those types of unsustainable resources the more we are depleting the source. Another example could be the harvesting of trees for firewood. If we take all of the trees at one time then there will be none left for the following year's firewood needs. Holmgren asked us a very useful question to help guide us on this understanding. 'Will the function or product which the resource is being used for last at least as long as it took nature to generate that resource?'[7]

Holmgren's exploration of this principle is still focused on biological resources despite the name change to renewables. He has, for example, a big preference on trees over solar as sources of energy, as solar has a high embodied energy and trees have many other uses and ecological benefits. This is technological resources versus biological resources, and putting a higher faith in nature-based solutions is the way forward. What I really like is the way he distinguishes between renewable resources and renewable services. For example, a tree carefully harvested for fuel and timber is seen as a renewable resource whereas a tree used for shade is a '*passive function*' or renewable service which has no impact upon the tree. Other examples of renewable services are the microbes turning the soil, worms processing compost and the bedrock filtering water. I am reminded of the principle, ***everything gardens*** as Holmgren describes this.

Permaculture's emphasis has always been on becoming a producer as a way to limit our consumer attitudes. In this way we can reduce our non-renewable resource use, use energy conservation methods, increase ***energy efficiency*** and become more familiar with the renewable resources around this. Why use a tumble drier when it's sunny and breezy outside? Why use air conditioning when there is the shade of a tree nearby?[8]

Appropriate technology

There are quite a few permaculture teachers' handouts and articles that include ***appropriate technology*** or ***alternative technology*** as one of the principles, including the article already mentioned by John Quinney and those by Sandy Cruz[9] and Patricia Allison.[10] The term appropriate technology evolved from work by Schumacher in his book *Small is Beautiful: A Study of Economics as if People Mattered*[11] and the campaigning by Mahatma Gandhi who advocated for small, local, village-based technology.[12] Appropriate technology is defined as hands on, people centred, ***small scale***, affordable and controlled by those that use it.

Typical applications include the use of hand powered or bicycle powered tools over petrol powered machines, or using very local hands-on resources for building, such as cob and straw. Empowerment of people, minimising fossil fuel use and thus helping the shift towards renewable energy are key reasons for the adoption of such technology. Examples include hand pumps, brick making devices or food processing technology, such as the universal nut sheller.[13] The nut sheller takes less than $50 to make, can be operated by hand or by bicycle and can shell up to 57kg of peanuts in an hour, thus overcoming the key ***limiting factor*** in making peanut production economically viable at a small scale. Appropriate technology encourages design accessibility with plans available for free for the construction of machines, which is also an application of the ***fair shares*** and ***people care*** ethics too, preventing a minority of people profiting from such machines.

I prefer the term appropriate technology over alternative technology as even though the latter includes renewables, such as solar and wind, they could be either smaller scale and thus appropriate or large scale and controlled by corporations. I got a sense of the grey area in between alternative and appropriate technologies when I worked out how to fix our old wood burning stove back boiler. It involved replacing parts, asking for some advice and using tools I had not used before but it was possible and I did it. If one of our solar panels breaks then I don't think I would have the skills or technology to fix it. Some components, such as solar panels, allow us to use the sun as a renewable resource but they are not made from renewable or easily fixable materials. There is a balance to be struck here when we are using non-renewables to allow us to access renewable energies and I think seeing how appropriate the technology is can be one of our tests.

Are renewables always best?

Sarah Spencer brought all of the resource related principles under one roof when she coined ***value local, biological and renewable resources***, as part of her Think like a Tree work.[14] Bringing the local in I think is key. One aspect of renewable energy is the vast scale that it can take. Massive wind and solar farms bring ***efficiency*** but also the powers of capitalism and in some cases a terrible environmental impact. Focusing on ***small scale*** local resources helps to maintain an ethical aspect through visibility and connection to the potential negative impacts. This approach should maximise ***integration*** and ***minimise waste***. It also brings us more control over the energy that we are trying to ***catch and store***. For example, there are some great community controlled renewable energy power systems on some of the Scottish islands, such as the Isle of Eigg.[15]

Whenever I need a resource I always ask if it is on site already or can we make it from something on site. If not then is there someone I know locally who already has it, and only after exhausting those possibilities will I consider buying that resource. I recently went through that process when needing spare rabbit netting

Using reclaimed tyres as membrane weights © Jenny Pickerell

to fix a perimeter fence in the market garden, resulting in no purchase of new material. This has been referred to as the *buyerarchy of needs* by Sarah Lasarovic.[16]

There are also times when I think we should not be using renewable resources and this is when there is an existing resource, renewable or not, already in use. For example, we already have old car tyres to hold down the weed barrier membranes in the veg garden. So why would we dispose of the tyres, even though they are non-renewable, and replace them with something renewable such as timber? Energy would be required to dispose of the tyres and to produce their replacement. There could also be pollution created by the disposal of tyres, depending on what happens to them next.

Once a plastic based item can no longer be reused or it can not be replaced with a second-hand option and it needs replacing, then consider making the switch to a more sustainable material. For example, switching to wooden toothbrushes and never buying a plastic one ever again or switching to compostable bags for lightweight items such as salad and phasing out plastic bags. Or maybe as an artist, switching to natural mediums such as charcoal and away from plastic materials and high tech options. In some cases the non-renewable item is in so much abundance I wonder if we will ever switch to the renewable version. For example, clothes at charity shops or online second-hand clothing platforms seem to never run out. Is it better to consume those instead of buying brand new items made from natural renewable materials?

Stepping stones

The Transition Town[17] movement is all about moving away from fossil fuels and shifting towards sustainable renewable resources. We need help unpeeling ourselves from the myriad of fossil fuels that are everywhere and in everything. This is where design processes can be handy, to observe each fossil fuel item in our lives and bit by bit, one ***small and slow*** step at a time, break away from them. For example, switching to solar and renewable power for your home by changing your energy provider or installing your own, decreasing energy consumption in your house and reducing plastic packaging. Here is a gift idea, get a Wonder Bag.[18] They are super insulated bags that can be used to slow cook food and keep food hot thus saving on cooking fuel.

Maybe instead you will start to grow your own food and save those precious renewable resources called seeds so that you can sow the same crops next year too. Maybe you will make compost or even build a compost toilet, giving you more access to the basic renewable resource of fertile soil. Or next time you need to buy an item you will choose the most sustainable item and one that is known to last a long time and rarely needs fixing.

More appropriately, in a different situation it might be that you need to review your money, time, focus and efforts, all of which are resources that we can manage and shift towards more sustainable practices.

Wonder bag © Wilf Richards

Fair shares

It is the increase in our net ***yields*** that generates excess resources. And when we have more than we need in resources then it is time to give away, trade or sell that excess. This is one of the aspects of the ***fair shares*** ethic, called ***share your surplus***. Another key aspect of fair shares is ***limits to consumption***, again particularly relevant to this chapter as we aim to limit our use of all resources. The only resource maybe we don't need to limit is creativity, as Starhawk once described it thus: ***creativity is an unlimited resource.***[19]

Putting limits on our consumption of renewable resources is part of that picture. There will be only so much wool from sheep in any given year and only so much wind for power generation and only so many trees that can be coppiced each winter. At the same time, to not harvest the wood through coppicing would mean a lower ***diversity*** of ground flora plants, and sheep need shearing. In a few cases limits can be relaxed. I know that we can take and eat as many rabbits and squirrels as can be provided to us by local hunters, as those animals can breed with far more speed than hunters have skill or motivation. So we need to combine this principle with the ***fair shares ethic***, put limits on greed and only take what we need and what the land can give. This may require policies, regulations and taboos, maybe manifesting as stories of caution and creation of sacred spaces as we see in indigenous cultures.

What constitutes a sustainable yield may only be shown after centuries of experience and mistakes. This knowledge then needs to become incorporated in a cultural tradition that can pass on the learnt wisdom.

David Holmgren[20]

Resilience

Written in collaboration with Aranya and Wenderlynn Bagnall

Is resilience a principle?

Permaculture does not explicitly have a principle about resilience and yet we talk about it a lot and many of our designs are clearly considering it. We utilise ***multiple functions and elements*** to increase resilience. We ***catch and store energy*** in times of abundance so that we can cope with times of scarcity. The earliest books about permaculture don't use the word resilience. Instead you will find descriptions of stability in natural systems maintained through checks and balances. The word resilience only started to gain traction and common use in the 1990s, probably as a reflection of the growing awareness of climate change and ecological collapse, and it certainly became a big buzzword with the emergence of the Transition Town movement from about 2005.

David Holmgren has dismissed the notion of resilience being a principle in its own right, as described in his article *Resilience and Reciprocity: Key Permaculture Concepts* written at some point after 2007.[1] In the article he explores how each of his twelve principles contributes to resilience. For example, we need to ***observe*** and keep a look out for possible dangers and changes to a system. By ***obtaining a yield*** from our own systems we can be more self-reliant and less likely to experience shocks in neighbouring systems. I can see the relevance of many of the principles to supporting a resilient system just as he describes, including ***catch and store***, awareness of feedback systems, the importance of ***renewable resources***, ***diversity***, ***integration*** and the ability to respond to change through ***adaptation***. Some of the connections he makes though are questionable for me. For example, I am not certain that ***small and slow*** solutions are going to be more resilient all of the time. There may be agility in the small but a larger size is very useful to cope with a shock because typically more resources are available for recovery.

Whether or not you regard resilience as a principle or a concept I felt the need to have a chapter about it. I see mutual connections between all of the principles, just as I have described in many of these chapters, so I don't see that as a reason to exclude it. We aim to achieve resilience with the systems we design and thus as a higher level goal it sounds like it could be a principle. We also know that due to the impermanence of the world, resilience is something we need to keep an eye

Think like a Tree natural principles

Group 1 Principles – Observation

Observe and interact
Be part of the natural world
Follow nature's patterns
Engage with the present
Learn from what's gone before
View the big picture then the detail
Listen to your internal ecosystem
Nature doesn't do straight lines

Group 2 Principles – Purpose

Live with purpose
Don't stop growing
Learn by doing
Good enough is good enough
Learn self-regulation
Nature doesn't do perfection
Catch and use energy effectively
Manage your energy not your time
Focus your energy where it can have the most effect
Design for maximum yields for minimum overall effort
Find slow and small solutions
Seek out 'flow' experiences
Change can happen in an instant
Value your uniqueness
Leave wild spaces in your life
Use your edge

Group 3 Principles – Surroundings

Create ideal conditions to thrive
Find your niche
Invest in your wellbeing
Plan for each need to be well supported
Each important need should be supported by several different strategies
Each activity or element should fulfil several needs
Feed your roots
Only feed what you want to grow
Waste nothing, recycle everything
Take no more than you need
Value local, biological and renewable resources

Group 4 Principles – Connection

Know that everything is connected
Everything works both ways
Value diversity
Cultivate co-operative relationships
Aggregate scattered elements into something greater
Integrate don't segregate
Pick your battles
Accept feedback
Do productive, not paralysing
Share the abundance

Group 5 Principles – Resilience

Develop resilience
Know that life is a struggle
Bend with the wind
Let it go where it wants to go
Creatively respond to change
Life moves
A bare space will always be filled
Remain flexible
Find solutions
Seek and offer support
Work from yourself outwards
Learn to heal yourself
Tune in to natural cycles
Take time to pause

Group 6 Principles – Future

Invest in the future
Know that death is part of life
Create conditions conducive to life
Replicate and build on strategies that work
Be a good ancestor
Strive to have enough, not more
Know that your actions can change the world
Love life!

Principles from the book Think like a Tree: the natural principles guide to life by Sarah Spencer
www.thinklikeatree.co.uk

Think like a Tree

Think like a Tree principles by Sarah Spencer

on as the goalposts keep moving. For the sake of survival I propose that we make resilience one of our guiding principles for all of our designs and I am not the first to propose that.

Starhawk has already proposed ***resilience is true security*** in her set of principles.[2] She highlights the importance of ***diversity*** and ***multiple elements*** to ensure backups in any system, the relevance of ***starting small and slow*** so that early mistakes can be made carefully, and the need to *design for catastrophe.* Sarah Spencer also includes a whole section about resilience, with multiple principles in her book *Think like a Tree,*[3] focusing on the ability to bounce back or bounce forward after a shock. And permaculture teacher Rakesh Rootsman Rak includes resilience in his teaching, focusing on the need for multiple connections of support and backup systems that provide options of redundancy if any one of those backups fail.[4]

What is resilience?

The word resilience comes from the Latin *resilio* meaning I spring back. It describes the ability to cope with shocks to a system. The shock will result in the system taking a load that, like a spring or an elastic band, might return to its original shape, lose its original shape or suddenly snap. The more resilient the system is the less likely it will lose that shape in the first place or will recover quickly and not snap. It is the ability to survive setbacks and resist external pressure to change and possibly even flourish in that adversity. We may get knocked down, but if we are resilient then we will just dust ourselves off and *get back on the horse.* The definition will shift depending on the context: we can consider resilience in an individual, a community, organisation, business, ecosystem and even the entire planet. Resilience has an overlap with ***adaptation*** but they are not the same. Adaptation refers to the changes that we make to better fit with the environment that we find ourselves in. It is normally a change for the better and the external changes may be positive or negative. Resilience normally focuses on our capacity to make changes in the face of adversity. The external factors are always negative in the case of resilience.

Humans are having a massive impact upon the resilience capacity of natural ecosystems through pollution, industry, agriculture, overfishing, habitat destruction and of course climate change. Changes to the land, water cycles and nutrient flows can all contribute to that loss of capacity and ability of an ecology to recover from the shock of intense human activity. We can see the resilience of a healthy piece of land when we are gardening it: if we stop managing it for a while, it will soon turn back into a jungle of thistles and brambles as it heads towards a ***succession*** of fantastic plant growth. But if our impact on the land is too high through pollution, such as chemical runoff from agriculture or deforestation, causing the loss of key species, that recovery may never happen or a new degraded stable state, such as desert, may be the result. In recent years we have seen the collapse of lake ecosystems through eutrophication and signs that coral reef systems are rapidly degrading due to changes in sea temperature and acidity.[5] Whenever we are considering developments, we need to consider the impacts we are having on the environment and always take into account the resilience of the local ecology. This is one of the aspects of the permaculture ethic of ***earth care***.

We know that some species have survived previous mass extinctions, such as crocodiles, alligators and flowering plants. And it was the early birds and mammals that survived when the dinosaurs became extinct, probably due to the ability to fly away or hide underground. These survivors are perfect examples of resilient species, although they may also simply be the lucky ones. Given a different crisis, such as the current climatic changes, those previous survivors will need to be lucky this time too.

Resilience is important

The need for resilience is crucial for survival. All systems and species, including ourselves, experience extreme weather, diseases, accidents, floods, fire and many other types of shocks. Shocks can happen at many different scales, from the individual and species level through to a whole habitat, society, planet and beyond. We are not immune to difficult situations, adversity does not discriminate. Humans have been particularly effective at creating situations of horror to test the resilience of individuals and cultures through war, trauma, abuse, dictatorships and terror. We can be affected by shocks in many different ways including at the organisational, physical, psychological, mental, emotional and spiritual levels. This is our first reason why resilience is important. We are not immune from difficulty and suffering. Accepting that shit happens is a great start.

If there is an inability to bounce back from a shock then, depending on the situation, the result could be collapse, failure or even death. It could be the mental breakdown of an individual, the bankruptcy of a business, the failure of a project, the extinction of a species or even the ecological collapse of the planet that may take millions of years to recover. This is our second reason to consider resilience in all of our designs. A lack of resilience could result in a premature end.

Hopefully initial shocks we experience are small and will give us the opportunity and lessons we need to learn. The development of this new-found wisdom and grit will enable us to put into place a more resilient system for the next time a shock is likely to knock at our door.

Building personal resilience

Dr Lucy Hone, a director of the New Zealand Institute of Wellbeing and Resilience, shares in her TED talk three key requirements to obtain personal resilience.[6] As I have already described, the first is to accept that shit happens. That way when suffering does occur it is not a surprise and we can avoid developing a victim mentality. Secondly, she says focus on things you can change and accept the things you can't change. Move from a place of fear to the positive opportunities of a crisis and recall the gratitude for what you have. You will get through this, hopefully. Lastly, she points out the need for reflection and to choose the options that will help you. Remember that you do have some control over your thoughts and actions.

The probability of some shocks in your life could be reduced through regular risk assessments. This could be through making plans or listening to your intuitive fear and worst-case scenario imagination. Now consider your capacity and resources to cope with that possible shock. Are you prepared and know who can help?

Some shocks are properly unexpected and very difficult to plan for. Consider the situation of an earthquake. Maybe you live in an earthquake-prone area and so you may already have a plan in place but the timing will be unexpected or at very short notice. The people affected may have lost their homes and loved ones. There will be a need for emergency aid, food and resources. It could feel like everything is lost, everything has been taken from you. The resilience of your mind might be the only thing that will get you through this crisis. Let's consider next what might be in your mental resilience toolkit.

Whatever the source of the shock is, the possible outcome is exhaustion or burnout. With experience you will recognise your symptoms of stress before you reach burnout point. I recommend carrying out a stress management design to explore your personal symptoms and indicators of stress. When your indicators appear it is always a good idea to stop, take a breath and sit down. Your breathing needs to be deep in the belly with longer breaths out than in, as this will activate your parasympathetic nervous system and bring calm to the mind, increasing the possibility of responding to the crisis in a clear and focused manner.

There are many other ways to maintain your wellbeing, such as considering your food, sleep, rest, recreation and exercise, all of which will support you in being more resilient. Healthy food and care of your gut microbiome will set you up for a strong immune system to fight off disease. A good quantity of sleep and rest will recharge your batteries, giving you the ability and capacity to cope with things when awake. Rest might include time alone, a recreational activity, hanging out with friends or having a celebration. Vigorous regular exercise, such as high-intensity interval training, yoga or running is great for grounding, reducing inflammation, improved mental health and increased energy levels. Further ways to minimise the likelihood of burnout include setting boundaries and limits, being assertive and knowing your capacity. It may include allowing time for chaos: I will frequently add ten-minute slots into a teaching plan to allow for the unexpected, or book in days of rest especially after big events to allow for recovery.

You can tell when we have developed skills of resilience as there are indicators that you can look out for, such as the ability to think strategically and confidently in a crisis rather than just react. You will also look out for the signs of stress and have in place plans to maintain your wellbeing. You are more likely to cope in a difficult situation, less likely to resort to substance abuse, less likely to suffer from trauma and more likely to recover from trauma quickly. Your emotional intelligence will be developed so that you don't jump to conclusions, don't judge quickly, have more empathy and an increased ability to discuss and control emotions. You will also be willing to evaluate the successes and failures of your systems and be willing to redesign or tweak a system if it fails rather than ignoring it and hoping the nightmare will go away.

This personal level of resilience is all about ***people care.*** When someone faces a disaster and doesn't cope well or burnout occurs, we need to also take care not to blame individuals. We all have a limit on our capacity and sometimes we need outside support. There are times when the system could be blamed rather than the individual. The situations that we live in can be a big test of one's resilience. Maybe the workplace is toxic or the government does not give enough support. Maybe it is the system that needs to change rather than just telling the individual to be more resilient. This is where working together with those that have a similar experience will help. We need to find that common ground with others so that mutual strength can be obtained to push for change in the systems we live in. If this does not exist then we may be forced into finding coping strategies to deal with the situation.

Developing community resilience

So maybe now you are prepared as an individual and feeling mentally resilient to any possibility. What about your community? Are you prepared as a group? We don't operate in isolation and we aren't ever completely self-sufficient. We need people around us. Acts of ***cooperation*** to build community will help everyone involved to prepare for and limit the likelihood of a shock. It will also reduce the vulnerability of any individual and provide mutual support to get you all through the inevitable crisis. Being part of a ***diverse*** community, not just lots of people like you, is important to increase resilience. Are you part of a community already? Maybe it is in your neighbourhood, a group of friends, a local institution or club.

In his book *Stand up Straight: 10 life Lessons from the Royal Military Academy Sandhurst*, Major General Paul Nansen talks frequently about resilience in the context of officer training.[7] At Sandhurst they push the cadets to the ***edge***, what they refer to as the *threshold of failure*. They then train them in positive thinking, optimism and reflective processes to learn from their failures and work out what went wrong. From there they are taught to identify their personal strengths and weaknesses, and thus to build ***cooperative*** teams that are mutually supportive, minimising the risk of future failure and building team resilience. Individuals need to know when another team member could do a better job than themselves, and when they need to rest to avoid burnout.

Once we have community, we need to maintain it, and here the relationships are key. It is easy to slip into thinking that a community is made up of individuals but it is the connections between those individuals that are essential. It can be easy to focus too much on head-based knowledge and action and not give time for reflection. I know this all too well from my work at Abundant Earth,[8] where we've had many crises but we always get through them by giving time to sit and shift towards a heart-based connection and giving emotions time. We all need processes and skills that enable connection between those involved. This could include empathy, compassionate communication skills and sharing difficulties with someone you trust.

A project will never fail because the land is poor,
it will fail because the people will fail to get along.
Patrick Whitefield[9]

Resilient projects and organisations

So we have personal resilience and maybe a resilient community, now what about the land we work, or the business we run or the project we manage. Communities are informal and social with many invisible relationships but projects and businesses are grounded in physical visible reality. What can we do to make these situations more resilient?

All organisations need contingency plans in place, to consider what will happen when change occurs, to future-proof against disaster. These plan Bs might be through the addition of extra ***elements*** in our systems so that the ***functions*** can continue to be fulfilled. This is commonly known as redundancy and has already been explained in detail in the functions and elements chapter. For example, we can have more ***diversity*** of food crops so that if one fails then many others hopefully will thrive. It might be through the additional harvesting of a key resource in times of abundance so that when scarcity strikes, such as drought, we know we have already ***caught and stored*** enough water to see us through.

Even though as an organisation Abundant Earth has considered resilience to external shocks in the design of its economic system, I am still surprised when we survive a shock, such as the economic impacts of the Covid pandemic. Our economic model is not only as a cooperative but also as an income-sharing system. This means that all income streams go to the same bank account and all directors receive the same wage independent of the amount of income that individual directors bring into the coop in that period. This provides both the business and the directors with financial stability. In comparison, anyone who is self-employed will recognise boom and bust cycles where income and expenditure vary through the seasons. Ideally you would ***catch and store*** your excess income for the lean times.

Self-Regulate

Written in collaboration with Beth Silverbirch

Regulate yourself

David Holmgren coined this principle as ***apply self-regulation and accept feedback*** in his book.[1] He clearly recommends that we need to discourage inappropriate behaviour in order to save this precious planet and ourselves. This principle is all about listening to feedback, looking out for lessons learnt and managing our behaviour. Failure to regulate ourselves could result in behaviour and actions that are detrimental to ourselves, others around us and our environment. If we are considering change we need to be aware of what we can change. According to the Stoic philosophy, the only things we control in our lives are our own thoughts and actions.[2] Trying to control other people's thoughts and actions normally leads to tension and conflict, therefore we should not try to control or be stressed about the thoughts and actions of others. Furthermore controlling your own thoughts seems pretty hard but we can, with practice, choose to decide how we respond to those thoughts.

Self-regulation is rarely mentioned in any permaculture text before Holmgren's book but it does get some cover in *A Designer's Manual*[3] in relation to the stability of and built-in feedback mechanisms in ecosystems. This was largely influenced by the Gaia hypothesis[4] and systems thinking. The Gaia hypothesis states that living organisms interact with their surroundings to create a whole Earth self-regulating system that maintains life on this planet. It refers to how organisms influence the stability of conditions here on Earth, such as global temperature, atmospheric oxygen levels and ocean salinity. It is a complex theory and those in the scientific community are still seeking further evidence to support it. This area of science has contributed a lot to our understanding of the importance of ecosystem feedback mechanisms and asks us to consider the dramatic effect humans are having on the whole planet, which could break those stable conditions, especially global temperatures.

This principle of self-regulation is probably one of the most important principles for us humans. It demands that we change our behaviour to prevent severe climate change and mass extinction. The combination of climate legislation and personal self-regulation is required.[5] We so badly need to listen to the feedback we receive in its many forms, whether it's our inner instinctual voice, the people around us or the earth itself. The feedback loop from the earth has been getting increasingly louder, and it is now sounding like a microphone pushed into the loudspeaker of a rock band. It is screaming stop, minimise, reduce, tweak and change.

There is a clear link between the principle of self-regulation and the permaculture ethics, which include the key messages of ***earth care***, ***people care*** and ***limits to consumption***. There is a further link to the original permaculture prime directive as stated by Bill Mollison: 'the only ethical decision is to take responsibility for our own existence and that of our children'.[6]

Reinforcing and balancing feedback mechanisms

Concepts about feedback systems exist in multiple disciplines, including biology, psychology, economics and electronics. There are two key types of feedback in all of these disciplines: positive feedback loops[7] and negative feedback loops.[8]

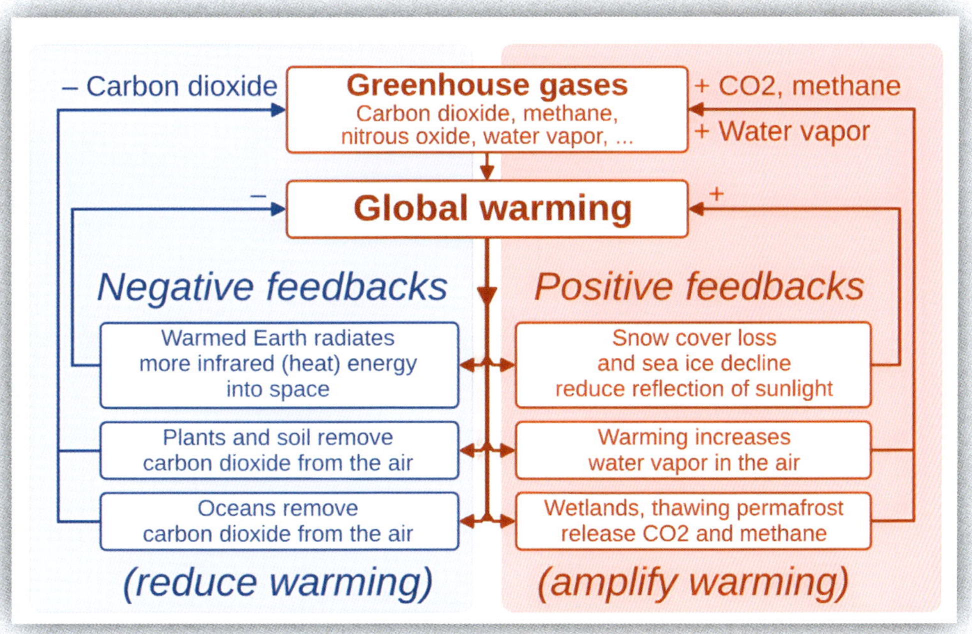

Global warming feedback loops © Wikipedia: RCraig09[9]

Positive feedback loops encourage more of the original input. Systems thinker Donella Meadows, preferred to refer to these as reinforcing feedback loops so as to avoid the notion that these loops could be judged as positive, as in 'good'. An example might be putting a microphone to a speaker which will create an infinite loop of noise getting louder and louder. It also applies to unchecked animal population growth, such as we have seen in humans in the last two centuries. Our explosion in numbers has been further supported by fossil fuel availability, technological developments and the growth of agriculture. We can also see reinforcing feedback loops in the escalating panic caused by a run on a bank. In biology there are also examples, such as hormones released in childbirth and blood clotting, the more a particular chemical is released in the body the more of that chemical is encouraged to form. Maybe we can also see it in a social media post that goes viral; the more people the post reaches the more likely it is to be shared. Donella Meadows described these feedback loops as examples of growth and explosion, potentially leading to collapse of systems. Extreme examples of positive feedback loops leading to collapse are rare. These vicious circles typically only exist briefly before a ***limiting factor*** will trigger a negative feedback loop which will bring the system back to stability.

Negative feedback loops, or balancing feedback loops, as Donella Meadows called them, have a balancing or checking effect thus restricting the original input. The more deer there are, the more there are for predators to eat, which might increase predator numbers who will then eat more deer and so keep the deer population in check. We can also see negative feedback loops in the use of a thermostat to control room temperature. As it gets hotter in the room the thermostat, at a key temperature, will turn the heating off. We can also see it in increased body temperature which can cause us to sweat, thus cooling the body down. It is these checks and balances that steer a system towards stability and equilibrium.

Global warming includes both reinforcing and balancing feedback loops but the reinforcing feedback loops are stronger at the moment. One of those reinforcing loops includes the melting of permafrost as the planet warms, leading to the release of methane which in turn causes more warming. Whereas on the balancing feedback side we see plant growth increase with warmer temperatures which in turn means plants absorb more carbon dioxide thus balancing the temperature increase.

The story of the Kaibab deer in Arizona contributed to our understanding of the complexity of these feedback systems.[10] In 1906, President Roosevelt created a game reserve in the Kaibab Plateau. This was intended to conserve and increase deer numbers but the result was the opposite. Key deer predators were killed, livestock in the area were removed and deer hunting was banned. Thus a reinforcing feedback loop started with increased grazing available to the deer, resulting in their population increasing dramatically over the next twenty years. This in turn led to a ***limiting factor*** and a negative feedback loop kicking in which was the reduction in availability of good grazing. The overgrazing situation resulted in the starvation of thousands of deer and a drop back close to their original numbers. Our understanding of complex systems and the different types of interventions has since developed, as already explored in the ***least change*** chapter.

Patterns of processing feedback

We can see from these examples above that feedback ***patterns*** and self-regulation of systems exist in nature. As modern humans, who are a bit detached from natural systems and with the power of fossil fuels behind us, we have developed a tendency to ignore feedback loops, minimise our self-regulation and instead developed notions of infinite economic growth. I think we have created an illusion that we do not need to listen to feedback and we do not need to self-regulate.

I recently had a conversation with a young lad who was expressing how much fun he had been having over the summer. Off to Ibiza for a party then to Italy for a holiday, going to festivals and generally having a good time. Lots of positive feelings as the need for fun had been fulfilled. Then he said that he also felt guilty about all the flying. I pointed out that the guilt is a form of feedback from his own conscience and that it is an indicator of unmet needs. Those needs might be care for the planet and integrity: aligning his actions with his thoughts. He has a choice to listen to that feedback from his inner self and decide how much flying he will do in future. We can find many different ways to meet our needs and aim to consider all needs. Maybe the young lad might choose a different strategy to meet his need for fun closer to home next time.

There are ***patterns*** for how humans process and accept feedback, for example the transtheoretical model developed in the late 1970s.[11] This model has been found to apply to just about every type of change in human behaviour. It includes five stages: *precontemplation, contemplation, preparation, action and maintenance.* The feedback at an early stage of change, where we may be in denial, anger or resistance defines the *precontemplation* stage, but awareness is building that change may be required. Next comes the *contemplation* that change is required, but not just now because there is still a need for or addiction to the current behaviour. At this stage we need to increase the pros for change and decrease the cons. Eventually the feedback being received might start to be accepted and *preparation* of options for change will begin to be considered and a plan of action may form. Finally feedback is totally accepted, advice might be sought and *action* towards change begins. Changes in our behaviour may be vulnerable and fragile leading to a reversal, or back to square one, and so *maintenance* of this new habit is required.

The duration of these stages of change depends on the context, the level of addiction, social factors, political interventions, support we receive and the willingness to listen to ongoing feedback loops. As an ex-smoker I can see these stages clearly from my teen years of not caring and not listening to the evidence through to gradually receiving feedback from my body as a form of persistent coughing and craving, then trying to quit for short periods of time before finally quitting for good and then maintaining that habit through a healthy hatred of the tobacco industry.

The Transition Towns movement was influenced by this pattern and used it to inform action plans to shift us away from our addiction to fossil fuels. That change is so much harder because many of our needs, such as movement, financial safety, warmth, shelter, food, adventure and connection are often met with strategies that use fossil fuels. Instead we need to break down that change into ***small and slow*** solutions and get creative about finding different strategies to meet needs so that bit by bit we are no longer dependent on fossil fuels.

Strengthening our feedback skills

So from here I see two skills required to enable an ease with self-regulation and behaviour change. The first is our feedback skills, the second is our capacity to change. Let's consider feedback skills first. I repeatedly see our cat fluff up whenever a dog comes near. It might dance and hiss and make moves towards the dog. It is clearly saying 'Stay away, I might attack and I am scared all at the same time.' I always hope the dog will accept the feedback or that the cat will get out of the way sharpish, otherwise it could be a matter of life and death, or a large vet bill. These are clear visible signs of feedback but other types of feedback could be invisible, hard to interpret or subtle. We all have blindspots on our behaviour and so if some feedback is invisible to us, how can we self-regulate? We don't know what we don't know and so sometimes we require feedback from others as constructive criticism to highlight possibilities for change, even if to begin with it triggers denial of the need for change.

Tuning into more subtle forms of feedback is tricky. Maybe we repeatedly get belly aches and are trying to work out the cause. Was it something you ate or an infection or a form of stress? Maybe it is a lower back pain that keeps repeating and you are wondering if you need to sit less or do more stretching. Or maybe we want to hear feedback from course participants to see what went well and what could have been done differently. We may need to listen carefully, listen to our gut instincts, collect data or test theories to find the best solution, and probably only after many failed attempts. Sometimes the ***patterns*** can only become visible once the ***details*** of the data are transformed into a graph.

In these situations our permaculture design skills can thrive as we may consider designing a feedback system. There are a whole wealth of tools out there to support such designs, including survey templates, reflective journaling and feedback forms. The key decisions are deciding what questions to ask and what to measure and monitor. Many years ago I created a survey especially for our veg box customers. One of the questions I asked was about portion sizes, especially as some previous customers had stated the reason for leaving was that they were wasting veg. There can be many reasons for wasting veg, including not having enough time to cook in our busy lives. The majority of responses to the survey though said that the portion sizes of certain veg were too big. Consequently those veg portion sizes were reduced and we introduced a micro box option for those that wanted to reduce their portions even further.

Do be careful though not to get carried away with trying to measure everything. There is a well known saying, *what gets measured gets managed.*[12] It sounds great but we don't want to measure and manage everything. We do, though, want to increase the likelihood of receiving feedback as quickly as possible in a form that we can understand so that we know what needs changing.

As well as finding ways to receive feedback we also need to be skilled in giving feedback. This could be particularly important when we have been hurt or we see others hurting themselves. In professions such as non-violent communication, feedback is precise and specific, aiming to limit blame or judgement and expressed from the speaker's perspective.[13] This involves expressing how you are feeling in relation to what someone has said or done and making clear requests. For example: "When you packed the veg boxes yesterday I was impressed you did it so quickly and at the same time I felt concerned about the accuracy as you did it all in less than three hours, when it normally takes a lot longer than that. This morning I received emails from four customers saying they did not receive their eggs. Could you double check the packing of eggs next time." Compare that feedback with "You were sloppy and made loads of mistakes." A further sensitivity is included which involves asking if the receiver is willing to listen to feedback in the first place, and afterwards asking how that feedback was received. Feedback given in this way, with care, is more likely to be accepted. It's only feedback when it's received as feedback. It could easily be received as an attack or criticism instead. This is further complicated by power dynamics between the individuals involved. This is a highly nuanced skill but when we get it right then we are more likely to see change happen. The process involves lots of compassion, empathy, honesty, listening skills, awareness and curiosity.

Increasing capacity to change

Self-regulation asks us to change our behaviour and listen to feedback, and as we have seen that can be complicated. The ability to self-regulate requires capacity and energy. There will be times when we do not have the energy to change our behaviour or maintain that change. Having compassion, kindness and gentleness for ourselves at these times is so important. We can acknowledge our desire for change and mourn our moments when we can't self-regulate and need to seek rest to recharge our batteries. We need to accept our limitations and acknowledge that it is normal to not be perfect all the time.

There are various models about the management of change and what factors are required to support change. One such model, which is credited to both Mary Lippitt and Timothy Knoster, states that six elements are required for change.[14] They include *vision, consensus, skills, incentives, resources* and an *action plan*. The model shows that if any one of the requirements is missing then this will be a ***limiting factor*** on the desired change. For example, without a *vision* we will have confusion as we will not know which direction we are heading in. Without *consensus* then your collaborators may sabotage the change, without enough *skill* to enact the change you will feel unsure and anxious about your ability to act, without *incentives*

Element	Element	Element	Element	Element	Element	Outcome
Vision	Consensus	Skills	Incentives	Resources	Action Plan	Success
	Consensus	Skills	Incentives	Resources	Action Plan	Confusion
Vision		Skills	Incentives	Resources	Action Plan	Sabotage
Vision	Consensus		Incentives	Resources	Action Plan	Anxiety
Vision	Consensus	Skills		Resources	Action Plan	Resistance
Vision	Consensus	Skills	Incentives		Action Plan	Frustration
Vision	Consensus	Skills	Incentives	Resources		False Start

Model of change map © Timothy Knoster and Mary Lippitt

you may experience resistance, without *resources* you will feel frustrated that you cannot progress. And without an *action plan* you might find yourself going around in circles or making false starts.

I have found this model an incredibly useful indicator of what is missing when we are trying to enable change in a system. I give a strong focus to the telltale criteria signs that are shown on the right of the table above. If you are in the process of change then ask how it is going and how the progress is feeling. So for example, if the response to that question is frustration about the progress of change then look at the resources that you require as they are probably missing.

Maintaining change

Once we have made the desired change, such as giving up eating ultra-processed food, then we need to know how to maintain that change. Depending on the context, ongoing monitoring or regular feedback might help. We could easily experience a setback in our behaviour, a return to old ways. Reminding ourselves of our values, ethics and the hard work of making the shift in the first place could help. Reminding ourselves of the benefits, such as the gain in ***resilience*** or the increased love in action of ourselves, others and the planet. Knowing that the relapse is just temporary will be helpful. After all, you made the change before, you can do it again, even in this world of impermanence. Self-regulation, as described by entrepreneur and motivational speaker Jim Rohn, involves a choice between two pains: *the pain of regret or the pain of discipline.*[15] The pain of regret may have arisen from the joy of freedom. We simply want to do whatever we want and stuff the consequences. But those consequences are the motivation to avoid sliding backwards. Which pain is worse: regret or discipline? Know that your change of behaviour could be a key signal to encourage others to copy the ***pattern*** of an example that you set.

Small and Slow

Written in collaboration with Liz Postlethwaite

The pace of modern life is full throttle

Do you ever get the sensation that life is going too quickly, that there is too much to do and not enough time? Ever get the feeling that the world is a big place full of big ideas and run by people with big pockets? The scale of our societies stretches across the globe making us feel that we are only tiny insignificant dots. The pace of our modern lives accelerates as the population grows and cities expand. The number of potential interactions increases and all fuelled up for maximum speed. Even our walking pace increases in the big cities.[1] As does the likelihood of the spread of disease. Many of us got a hint of what a slower and smaller life could look like during the Covid pandemic, and maybe felt drawn towards it. I know some miss it, some of us enjoyed it, and some were deeply distressed by it. We eventually went back to normal, whether we liked it or not, and went full throttle once again. Of course some didn't stop working at all. Here in the green movement we always praise the small and the slow but this is in contradiction to the big fast motorway lane attitude of the modern world.

David Holmgren added ***use small and slow solutions*** to his twelve principles.[2] He promotes it as the principle to help us maintain systems and better manage our resources, both key steps towards sustainability. Smaller systems are less stressful and more within our capacities. They also encourage ***diversity*** as lots of small systems will display differences of activity and approach as compared to a few big systems, and that adds to the ***resilience*** of the whole system.

Prior to this Bill Mollison and Reny Mia Slay had put forward ***small scale intensive systems*** as a principle.[3] The focus though was the intensive part as the details explored both ***plant stacking*** and ***time stacking*** to maximise outputs within a small area that is manageable. Bill also asked us to *start at our back door*, get those things right and then everything else will be right too.[4] The principle has been tweaked in various ways over the years, such as Rosemary Morrow's request ***to start small and learn from change***[5] and Colleen Stevenson asking us to ***slow down and observe***.[6]

Slow and steady wins the race... sometimes

Many of us will be familiar with the story of the hare and the tortoise, one of Aesop's fables from the times of ancient Greece around 600 BC.[7] Our modern interpretation is that slow and steady wins the race but we know that is not always the case. In fact, to the ancient Greeks the emphasis was on the arrogance, idleness and overconfidence of the hare rather than the steadfastness of the tortoise. There are other variations of the story too, including those that involve trickery and sabotage to fix the race, such as the beautifully illustrated version by Helen Ward.[8] This additional angle is important to those that favour slow because it implies that the slow will only succeed if the fast fail. That is a hope with unsure foundations but we still sense that consistency and patience are better than rushing a job carelessly.

Most of us start from a position in life with limited resources, skills and time and so we tend towards a snail's pace to begin with anyway. There is also something restful, easy and relaxing when we can deliberately go slowly, and something magical when we can be patient and creatively wait for that best decision moment. Slowing down gives us time to focus on our creativity, when our best ideas might emerge – to move past the initial 'good' ideas to something much more exciting and dynamic. If you are able to include times of rest, a meditation habit or blatant idleness, then they can be great ways to lower the blood pressure, put the brakes on your busyness and lead to more fun and connection. It's time to slow down and get with the pace of the trees that grow imperceptibly slowly. Go poke a fire, read a book, have a pondering walk, do some doodling in the margins or mess about on boats. Take idleness seriously.[9] To truly slow down takes a serious commitment, which most of us fall short of due to societal pressures that focus upon work and productivity. I know I fail on that one on a regular basis with a workaholic tendency that needs curbing.

There is also a joy in slow food with diverse wholesome local ingredients, and eating your food slowly is great for gut health. The Slow Food movement began in Italy in the 1980s with a focus on celebrating local food cultures, the artisan and the traditional, that have been threatened with the rise of fast food culture.[10] A development from slow food has been the slow town movement, called Cittaslow, focused on better quality of life and resisting the effect that globalisation has on towns.[11] We all enjoy going to a town with lots of unique local shops rather than the uniformity of a city with only multinational corporations.

Slow has gained a bit of a buzz. There is now slow art,[12] slow radio[13] and slow tv,[14] all with the intention of slowing our busy lives down and expanding our reduced concentration spans. I was drawn to these newer aspects of the slow movement whilst researching about and creating a design to bring more rest into my life.[15] There are artists deliberately creating works of art that are in constant flow and change or require viewer interaction. These, such as work by video sculptor Gabriel Barcia Colombo,[16] require the viewer to stop and participate rather than take a quick look, a snapshot and then move on to the next piece.

A project, product or service may be *fast, cheap* or *good* but, it is said, it can only ever be two of those at the most and never all three. In fact, in trying to do all three you will soon find weakness in one aspect. A worse scenario is that we only achieve one of these aspects, for example, a building being put up fast but also being expensive and poor quality. The convenient popular products and services on the market lure us in with their offer of speed and low price. Well guess what they are lacking: avoiding the option of good quality should only happen in an emergency situation that requires a temporary bodge, or if we do not have the time or money to invest right now and the work is essential. I can think of a few times where it was necessary to throw a cheap tarp over something leaking until funds allowed a proper repair. At Abundant Earth our entire business development has been slow with a steady growth and many things done by hand. We have built everything bit by bit as profits have allowed, so hopefully anyone visiting can see things are good quality, have a degree of longevity and we know they have been relatively cheap. This triad of limited options of *fast, cheap and good* is sometimes known as the Builder's Triangle, possibly originating in 1950s project management theory.[17]

Many projects need to be slow. For any of you involved in community participation then you will know that a slow steady pace is required to bring people together, to build teams, to make careful decisions that involve everyone, enable relationships to develop with care and minimise conflict. One of the key functions of the process of design is to slow things down, giving us time to observe, gather data, research, analyse and make the best decisions.

Building slowness into our designs can be critically important too. Take for example flood protection and prevention. If we concentrate on ways to slow down the flow of water and thus prevent flooding we know that we will also be limiting

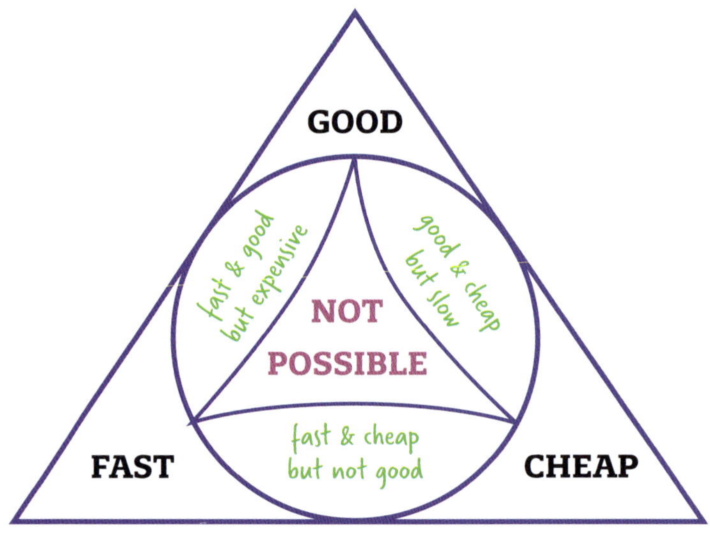

The builder's triangle

erosion and the loss of soil nutrients. There are many ways to do this, such as dams, leaky ponds and swales. These can ***catch and store*** the abundance of water for drought times and encourage plant growth by keeping the soil moist.

On a personal level we also want to avoid burnout, a common symptom of going too fast too quickly. When we are able to do things slowly this allows for the development of a solid foundation. This gives us time to reflect and learn the lessons from the feedback of what is working and what is not. If you are struggling to slow down then have a review of how you currently spend your time, and consider how you could liberate some of it by removing those things that do not serve you. If you are able to do this it can be great for your mental health and emotional capacity.

Small is beautiful

In 1973, E. F. Schumacher published his *Small is Beautiful* book[18] about economics, culture and resources, putting forward a case in favour of small-scale sustainable society as an alternative to the fossil-fuel-hungry globalisation. The book was largely inspired by the work of Leopold Kohr,[19] a philosophical anarchist and economist who mentored the Welsh independence movement and supported regional autonomy. Permaculture has been inspired by these strategies of working at the human scale and up to the level of the bioregion. Smaller systems are easier to manage, have fewer components, are less complex, can be sustained for longer and we are more likely to avoid burnout. Smaller systems that are commonly developed by permaculture designers include local networks, small farms, community gardens and neighbourhood-level projects. For some people these community level projects can be overwhelming. For many people it is helpful to start with just ourselves or our home. An allotment or growing some food at home may feel like a much more realistic option. Doing things small limits the potential negative impacts on the planet too.

Sure I'm a crank, a small element in a machine that makes revolutions.

E. F. Schumacher[20]

When we are faced with the need for change, small incremental changes, when possible, are the better choice. This allows time for ***adaptation*** and better decisions or the development of new habits, for example, changing our behaviour to improve our mental health. For someone trying to get over anxiety, it may be easier to identify the smallest fear first and then build confidence gradually in dealing with something bigger.

There is also the matter of scale and perspective. What is big to you may be small to me and vice versa, and the same applies to the scales of slow and fast too of course. For example, we love working at the local level, but is it just those hills, valleys and streams in our neighbourhood, or does it include that really big river and town 30 miles away? If you live in a really big country like the USA then that probably sounds very small to you.

Small is not just beautiful but also successful. The most successful species on the planet are those tiny invisible ones such as bacteria, viruses and fungi. It certainly is not us humans, unless we turn things around pretty quickly. Bacteria outweigh us, they certainly outnumber us, some of them can outlive us and they have been around on the planet the longest. Many would argue that humans are the most successful but I would say don't confuse success with dominance.

Start small then scale up

I like to phrase this principle as ***start small and slow*** with the emphasis on the word start. As we test an idea we would be foolish to invest all of our resources immediately. We are sensible to have doubts and caution and to use the art of research and development to test ideas. Once tested and the problems are ironed out then you should have a manageable project with good ***yields***. Know that even if you made mistakes at these early stages, you did it safely and within your means. Only then consider scaling up if you so desire. Knowing the boundaries of our projects gives us an ***edge*** to work from. From those edges we can consider expanding outwards to potentially increase our yields.

What is holding you back? Is it the lack of capacity, time, money, resources or confidence? A great model to indicate readiness to scale up or down is to use the flow state test developed by the psychologist Mihály Csíkszentmihályi in 1970.[21] Are you feeling anxious and overwhelmed by your task? In which case scale down or go steady. Or are you feeling bored and lacking stimulation? In which case scale up and increase the challenge. Finding the optimum point is called the flow state, where we effortlessly do. The opposite image shows boredom in the lower half where the challenge is low and the skill level is too high for the task, whereas the upper section shows the anxiety region where the challenge is high and the skill level is too low. We want to aim for a balance between our skill level and the level of the challenge that we face. We can use these feelings of anxiety and boredom as indicators of where we are on this map, and as a source of ***feedback*** to indicate the direction we need to head in.

Don't be put off by phrases such as *the bigger they are the harder they fall* as that really only applies to the truly powerful. That's not you, sorry. Instead when you sense the need and desire to scale up then do it carefully and use your design skills to do it. Identify what it is that is holding you back, what further skills and capacity do you need? Then remove those ***limiting factors*** and allow the system to grow to its new desirable manageable size. On several occasions over its twenty years of existence, we have scaled up the veg box scheme at Abundant Earth. These are tiny incremental changes from originally 15 households per week to 60 in 2023; simple shifts such as tweaks to admin systems, improvements in accounting, speed and

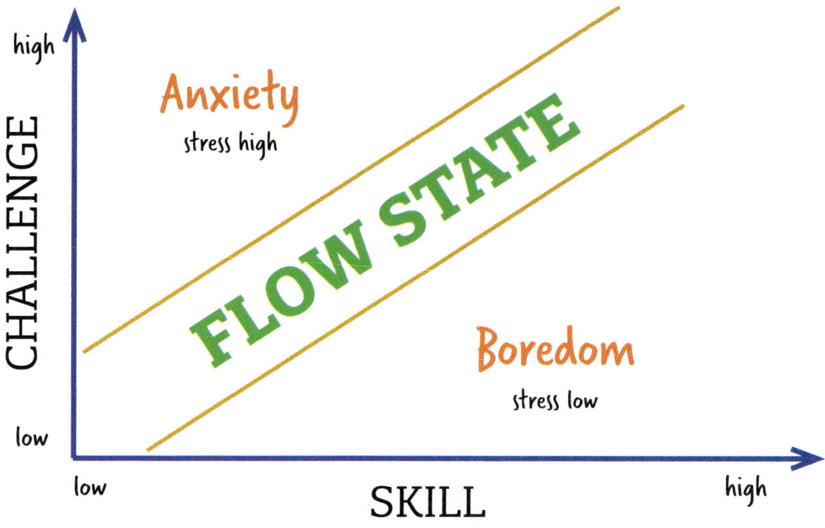

Flow state map

confidence in packing and rearrangements of packing space. With each tweak we have shifted the scale of the operation and added ***efficiency***. At every stage I consider what is now manageable and where the new boundary will be.

Maybe big can be beautiful too. There is evidence that species have a tendency to increase in size over the aeons of evolution as the bigger individuals of just about any given species are more likely to survive as they have more resources to cope with lean times.[22] There are of course ***limiting factors*** to these increases and counterbalancing forces. A bigger business may provide a more reliable source of income, create more employment, potentially bring ***diversity*** into the project and have ***efficiency*** benefits of a larger scale. It could also develop a larger capacity, build more connections and have more resources to cope with a crisis and therefore be more ***resilient*** to survive a shock to the system. You can always have the opportunity to carefully scale back if you need to.

The caution I have is that bigger and better could get greedy and I imagine it's even addictive when the money starts to flow. Small can be and should be better but as humans we are drawn and addicted to the big and bold as that has money, excitement and power. When Covid hit, Abundant Earth was inundated with requests from people to join our veg box scheme but we resisted, stuck to our maximum household numbers per week and continued to add people to our waiting list. I guessed that these potential new customers would not be with us for the long run, would put unnecessary stress upon us, trigger rapid changes that we could not manage and risk the collapse of the entire system once lockdown ceased.

Ensuring the ***ethics*** continue to be embedded as you scale up is also a challenge. An increased workforce is likely to result in a compromise in ***people care***. A higher number of products and services will impact ***earth care*** as more energy will be used. An increase in the centralisation and growing scale of a project could have an effect on ***fair shares***. Keeping an eye on the potential to maintain and even increase the ethical aspect is important as you scale up.

There are also times when we feel the need to go bigger and faster, like when the whole planet and all of its residents are suffering from climate chaos. Maybe our combined efforts of thousands of tiny slow projects all add up to one big fast change at some scale. It would be good to see some more positive shifts in my lifetime but this ain't a Hollywood movie and we may never know how this crisis unfolds.

The times are urgent: let's slow down.

Bayo Akomolafe[23]

The need for speed

The small and slow approach takes pressure off us, it can feel nice and gentle but with ecological collapse lapping at our doorstep the time is ripe for big solutions done quickly. Speed is of the essence now and the time for small and slow solutions to this massive problem has long gone, as far as I can tell. Permaculture teacher Russ Grayson said just the same in an article written in 2015, challenging this principle and telling us that small and slow may sometimes need to give way to big and fast. *'Principles and ideas, like all ideas and practices, should be questioned so as to introduce the changes that would keep them relevant to the conditions of a rapidly changing world. An out-of-date system is a system of no value at all.'*[24]

Just in case you hadn't noticed, that climate change thing is looking bad, very bad, we keep discovering it is a lot worse than the forecasts. It is unlikely that permaculture or any movement is mature enough to offer much to the vast scale and speed required now to resolve the climate change issue but do keep doing those amazing things that you do. It might just all add up and contribute, even at the snail's pace that we tend towards.

Let's not be scared of speeding up when we have the capacity, skills and resources available. We already like ***accelerating succession and evolution*** through tweaks to microclimates and planting of appropriate perennials. There are plenty of fast things in nature from cheetahs to the speed of light. We certainly could do with some of our ideas going viral and spreading like wildfire. Let us not be slow all of the time like an anemone waiting for its food to just float on by. Even Jeremy Clarkson understands that "speed has never killed anyone, suddenly becoming stationary, that's what gets you".[25]

There are times when we need to be quick. When there is an emergency, we need to respond. We would not expect the emergency services to be slow. One of the

Colleen Stevenson's beautiful principles drawing © Colleen Stevenson

Think like a Tree[26] principles is ***change can happen in an instant***. In our context that is likely to be a fire or a flood. Design skills give us the opportunity to anticipate those moments when a fast response is required. We can put contingency plans in place to be prepared for the emergency and even rehearse our responses. So know when you need to go slow and when you need to go fast, when you need to be small and when you need to be bigger, but get your skates on because death is just around the corner.

Stacking and Succession

Written in collaboration with Chris Warburton-Brown, Tomas Remiarz and Chris Evans

The stacking of plants

There is no direct mention of the term stacking in the first permaculture book but plenty of examples of very diverse perennial planting, such as intensive fruit and nut systems. The term ***stacking*** first appears in Bill Mollison's book *Permaculture Two: Practical Design for Town and Country in Permanent Agriculture* (1979).[1] There is a brief section titled the *stacking of plants,* clearly referring to the use of vertical space with multi-tiered highly ***diverse*** perennial plantings. The book demonstrates that there is clearly an art to the planting to ensure the plants thrive whilst considering the competition for light, nutrients and water. Spacing is key to ensure a balance between density of planting and productivity, and an understanding of each plant's particular requirements and strengths so as to place each plant in the best location and in relation to other plants is also needed. Each plant has its niche; we need to know it and work with that knowledge. The book even expresses the hope that such systems could be more productive and yet just as stable as a natural ecosystem.

Ecological succession

This stacking of plants comes directly from observations of ecological succession[2] of land. When left to its own devices we see a landscape gradually shift and change, for example from open field to scrub to woodland. With each shift new species appear and previous species move on or die out. The shift is always towards increased ***diversity***, increased variation in height of species and towards a climax landscape, as long as there are no ***limiting factors***, such as adverse climate, the availability of certain nutrients, low resources or particular species missing. Once the canopy closes then diversity might start to decrease as it may become too shady at the ground level for some plants to grow. This brings us back to the art of stacking plants. To maintain that diversity we need to get the light levels just right and so we may need to prune, thin or coppice some of the trees.

Woodland floor greening up before the leaves on the trees emerge © Jo Holleran

The details of the succession will vary according to climate, soil type and species within and near to the location. The climax state of a landscape is not the end of the journey as ecologies are never static and change is inevitable. That change might be as a disruption in the pathway of succession, such as new clearings in a woodland appearing as trees fall. This might be due to factors such as fire, wind, disease or herbivores grazing and so the cycle begins again, is knocked back a phase or even goes sideways, but will once again head towards increased layers stacked together. Some plants may remain dormant as seeds for years waiting for a clearing to appear.

This energy flow in nature gives us the opportunity to be ***working with nature*** rather than against it. A landscape could easily include many stages of succession within a relatively small space, creating a mosaic and patchwork of different habitats. This could be influenced by historical activities or grazing animals. We may find opportunities to ***cooperate*** with such livestock or may need to exclude them. Toby Hemenway asked us to ***collaborate with succession.***[3]

In examples of ecological succession, we can see plants thriving as they find their niche in space and time. Shade tolerant plants will be at the base of trees, whilst alders and willows thrive in the wetter soils. Plants that require higher light levels, such as ramsons and bluebells, will spring up earlier and flower before the leaves burst out on the trees above them. The woodland becomes green from the floor upwards during those spring months.

Succession can also refer to social systems, such as groups and businesses. It might be initiated by a pioneer, developing the ideal conditions to thrive. It may go through jumps in its development as capacity and resources build. Complexity may increase, maybe as income increases or the workforce expands, requiring new strategies, admin systems or managerial approaches. Abundant Earth reached such a jump recently when it was required to register for VAT. When our projects become mature they could be seen to be like a climax forest. We need to ensure that there are younger saplings, such as interns, apprentices and junior managers, stacked up underneath the leaders to take their place when they 'fall', creating a social clearing. These social clearings could include a manager moving onto a new job or even a death. Such clearings could provide the opportunity for the next generation to keep a system going or for new systems to emerge.

The term succession is also being used to describe the development of our gut microbiomes, from early pioneer species of beneficial bacteria through to established lower colon microbes to digest complex fibres. If all goes well then by the age of three we can have a fully developed 'forest community' of microbes down there helping us out. We just need to watch out for the microbial disruptors such as antibiotics and ultra-processed foods.[4]

Intensive and accelerating

In *Introduction to Permaculture*[5] Mollison and Slay consolidated and wrote about two connected principles. These are ***small-scale intensive systems: plant stacking and time stacking***, and also ***accelerating succession and evolution***.

The principle of ***small scale intensive systems*** started to shift the definition of the word stacking to ensure it included both time and space. The addition of the time aspect could include multiple people sharing land whereby many partners could utilise the same space but at differing times of the year. This could maximise ***functions*** and purposes and potentially increase the overall ***yield*** on that land. It could also include the early growing of annuals at the base of perennials whilst waiting for the perennials to develop. I also see this as common practice amongst commercial farmers whereby grassland is leased to livestock owners until spring time, then the land is left to grow grass which will be cut for hay in summer, with livestock returning in the autumn. This principle also included the definition that we are already familiar with, that of the ***efficient*** layering and stacking of plants through the knowledge of their niche requirements.

The principle of ***accelerating succession and evolution*** is quite different but connected. Rather than waiting for leaves to fall off trees to add fertility to the soil, we could accelerate that process by adding mulch and compost to the soil. Rather than waiting for a woodland to appear, we could speed up the process by planting trees or removing herbivores and allowing natural regeneration of trees.

Two years after removing the sheep in 2003 © Wilf Richards

The same area in 2019 © Wilf Richards

We may observe that the naturally regenerated trees grow faster and more successfully than the ones we plant. Any thriving we observe is a sign that the plant is occupying its optimum niche in time and space, supported by the right soil and microbial conditions. We could add plants with deep tap roots to bring up nutrients from sub soils, or add leguminous plants with their symbiotic microbes to fix nitrogen in the soil. We can see what thrives and what the land needs, and give it energy to shift towards a stacked, ***diverse***, stable system. This also clearly overlaps with ***working with nature***, creating the right conditions for fertile growth and strong roots. Once we have accelerated succession to the phase we require, we may become the disruptors to hold succession at that stage. This might be through pruning, coppicing and harvesting of produce. At Abundant Earth we removed sheep access to an area of our land shortly after we bought the land. It was a mix of grass and scrub to begin with but now after twenty years it is a woodland. It has gone through many stages of succession including gorse and brambles, then pioneer trees such as birch and hazel reestablishing themselves, and finally oak trees are gaining the upper hand and the gorse has become all leggy. Although this principle of accelerating succession is originally only applied to land-based settings we can also apply it to social settings such as developing your creative practice. As an artist, performer or musician, you will accelerate your development if you have a strong community of support around you, such as a fan base and fellow musicians.

Forest gardens

The principles of stacking and succession are most commonly applied to forest garden design. The term 'forest gardening' was first coined by Robert Hart[6] in the 1980s.[7] Robert pioneered forest gardens for temperate climates when he developed his multi-layered tree based cropping system in Shropshire. Forest gardens already existed though in tropical climates, such as Kerala, India. Robert Hart had been inspired by these examples and the work of Japanese peace activist, Toyohiko Kagawa.[8] Kagawa persuaded many Japanese upland farmers during the 1930s that the solution to their soil erosion problem lay in widespread tree-planting, and to ensure these trees were of the type that could also provide a harvest. This included walnuts that could be fed to pigs. Kagawa had been inspired by the book *Tree Crops: A Perennial Agriculture* (1929)[9] by Joseph Russell Smith,[10] who is regarded as the founder of agroforestry. Smith's main focus, as a geographer, was the planting of trees to prevent soil erosion especially in upland mountainous regions. This was extended to include food-cropping trees to add additional benefits to farmers.

There are many forms of agroforestry[11] today and they all involve at least two plants stacked together to increase ***yield*** on any given bit of land. The layers of a forest garden, although initially regarded as three, has expanded over the years to at least seven to include the herbs and climbers. The development of these successive layers is supported through the use of pioneer plants. These pioneers prepare and improve the conditions for the next phase. For example, gorse will shade the soil, protect young trees from herbivores and add nitrogen. Meanwhile other plants may bring up nutrients from the depths to the surface, create moisture-retaining mulch through leaf litter or tweak microclimatic conditions favourable to the next phase.

Forest gardens can help to conserve the soil, increase biodiversity, supply food, fuel and medicines for people and provide fodder for animals. They are full of multiple ***functions***, follow the energy of ecological succession, as well as being a system of stacked plants. They are multi-dimensional polycultures of various species all mixed up with various heights, niches and harvests at varying times. A mature forest garden feels like a Garden of Eden.

Is stacking also multitasking?

I have already put forward a case in the ***functions and elements*** chapter about the confusion between multiple functions and stacking. For me, the use of the term stacking to refer to multiple functions, purposes, goals or actions is permaculture slang. I have not been able to find the origin of this change of use but I do know that I am guilty of using this slang myself. I try to avoid using the term stacking in this way to help people new to permaculture not get confused between the two. I think they are different as I hope you can now appreciate. Stacking is more about visible physical properties of niche, space, time and placement in contrast to multiple functions which are often invisible, such as aims and criteria. But inevitably you will see and hear the term stacking to refer to both visible and invisible aspects. Let's see what that looks like.

Let's say, rather than going to the shops in the car to pick up one item, instead we consider what else we could do whilst out and about before jumping in the car. We can look for the possibility of *stacking* in additional tasks. Instead of immediately going to get those screws we need for a job, we wait until later, until we have built up a list of items needed. That patient waiting now means one ***efficient*** trip, which also minimises our ***waste.*** We could also ask others around us if they need anything as well and add in some ***people care*** whilst we're at it. Maybe we even decide to go on our bike, to add ***earth care*** to the mix. Now that I think about it, maybe I can stack in some more examples to explain this situation!

As well as physical multitasking we could also refer to stacking in an abstract way as multiple purposes or goals. I think it is rare to see any given action as only having one purpose or goal but maybe that is because I am a permaculture designer. The corporate focus can be very much on the money, this can literally be the sole purpose for some businesses, that and maybe power. Whenever we are considering a permaculture design though, we will be creating a list of goals. That might include food, education, biodiversity, livelihood and wellbeing. Typically our goals will link to all three ethics as well and in this way stacking purposes can be transformative.

Be warned: you can over-stack

So next thing I know, I have over-stacked the trip to the shops. I suddenly have dozens of things I need to do, too many tasks to carry out and all within a tiny time frame. I am feeling stressed and need to balance the stacking of jobs with the principle ***small and slow***.

Equally in a forest garden setting, if we plant too densely then one plant will struggle for light and others may dominate. We may have over-stacked or over-stocked our garden and may need to remove some plants casting too much shade. We need to be careful to not ***over-integrate*** our plants and create unnecessary competition for natural resources such as light or nutrients. As with many principles, balance is required.

Work with Nature

Written in collaboration with Jo Holleran

The good life

Whilst a growing number of folk are determined to live lightly to reduce their personal impact upon this precious earth and reverse the damage done, the majority of us continue to be both exploitative and degenerative. We know we are not working with nature. We can see with our own eyes our destruction and abuse of nature. We leave rubbish lying around and pollute rivers, we use products made from fossil fuels and burn that ancient sunlight at an alarming rate. We see the loss of habitats and species, we travel vast distances and build powerful machines to dig and destroy. We can see the evidence of our lack of care everywhere.

Hands in the soil © Walter Lewis taken at Abundant Earth[1]

We know and many generations before us have known that we just need to be more natural, have some nature connection, live a natural life, get back to the land, live in a mud hut, grow some veggies, go native, be green, become sustainable, find our Indigenous roots, learn from nature, imitate nature, somehow work with nature, be nature.

Bill Mollison first mentions the ***working with nature*** principle as the underlying philosophy of permaculture in his book *Permaculture Two.*[2] He clearly gives credit to Masanobu Fukuoka[3] and his book *One Straw Revolution,*[4] which outlines the natural farming philosophy.

> *Perhaps Fukuoka, in his book The One Straw Revolution, has best stated the basic philosophy of permaculture. In brief, it is a philosophy of* ***working with, rather than against nature****; of protracted and thoughtful observation rather than protracted and thoughtless labour; and of looking at plants and animals in all their functions, rather than treating any area as a single-product system. The difference is like that which exists between the Aboriginal and the ploughman: the latter is seen as one who would cut open his mother's breast to obtain milk, the former takes only what is given freely, and takes it with due reverence.*
>
> **Bill Mollison**[5]

By the time we get to the *Designer's Manual,*[6] Bill clearly states ***work with nature, rather than against it*** as one of the design principles. This is a principle that has gone largely unchanged since first being proposed. He talks about assisting rather impeding natural elements and forces. He refers to ecological ***succession*** of the land, working with that energy to provide microclimatic conditions for planting perennials and dismisses the use of pesticides as an example of putting nature out of balance.

So just what is nature?

When we look at definitions of the word nature it normally involves the exclusion of humans. It is often defined as something outside of us. It is typically all of the plants, animals and all other life along with the water, earth and atmosphere. A mix of biology, physics, chemistry, geography and geology. It normally does not include anything made by people or the people themselves. This is an error and a perception that we need to change.

We are part of nature, we are nature. We have become separated from it and it is this cut that needs healing. Once we can connect and be nature then we can fully care again. We are not isolated from our environment. Whatever we do to the earth, we do to ourselves. Nature is a complex chaotic mash of systems and interconnected beings both organic and inorganic. It is all alive, even the rivers and rocks, not just the biological entities.

Leaves fall and decay. Twigs of trees are broken off by a strong wind. The trunks of trees are blown down and lie on the earth. Animals and plants begin to eat this organic matter. These tiny animals and plants also die and decay, and become part of the soil, along with other organic wastes. Soon earthworms and ants eat these piles of rotting matter and they themselves return to the earth. This process of decay is the beginning of birth, part of the natural circle of life. The word 'nature' means this circle.

Hideki Inoue[7]

Nature is not a machine with separate parts and components, and we are not one of those parts. The microbes that occupy our guts also cover our skin and are in the food we eat and soil from which that food comes.

How do we work with nature?

If we are nature then how do we work with it? Permaculture designers love creating ***integrated*** systems, involving multiple ***elements*** working together with lots of beneficial relationships and high ***yields***, where our work is minimised and where we ***produce no waste***. This, at least at a ***pattern*** level, is how we work with nature.

Can you see the ridge and furrow? © Wilf Richards

First of all we need to shift our perspective and understand more about what nature is. Then we need to connect to it. This could be ***observing***, exploring and learning to read the landscape. I highly recommend Patrick Whitefield's book[8] on this subject. It could be understanding your local geology, geography and biology. It could be understanding the shape of your land, the slopes, aspects and altitudes, and how these influence what lives there. What else lives in your neighbourhood? What is the weather like? How does it change through the seasons? Do you have seasons where you live? Do you change as the seasons change? See how the landscape changes through time, and get an understanding of how it has changed over the years. I love pointing out the ridge and furrow to visitors on our land which was created by medieval farmers.

You might explore the water on your site, the rainfall, rivers and streams, the ice and frosts. I love seeing how the course of our local river has shifted over the last twenty years and then looking at the historical maps and seeing how it has changed since the early 1800s. I can also imagine how the river carved out the valley since the end of the last ice age.

It might also include spending time trying to identify the various species in your area or working out the various microclimatic variations from that sheltered sunny spot through to the exposed windy location. You might see how the land has been through ecological ***succession*** from grassland to scrub to woodland, or explore the soil by digging or learning about the plants who can tell you about the condition of the soil. Maybe you will see how everything is cyclic, or as Starhawk put it ***nature moves in circles.***[9] You will see how everything is connected and how death is essential for life. Maybe you will feel that connection now and let any notion of separation from nature disappear.

Learning the gentle art of avoidance and how to live with nature

In France, where Jo Holleran lives, her slice of hillside is densely populated with a huge diversity of species. How do they co-exist without war breaking out? The answer seems to be the practice of the gentle art of avoidance. Badgers stop to lay their scent every few steps, foxes travel with their noses to the ground, squirrels fly through the canopy, golden orioles build their nest in the same tree year after year, male roe deer rub their antlers on bark as they pass by. All senses are actively employed to pick up on these hints and to determine what has gone before and when. When different species accidentally encounter each other at proximity, her trail camera footage suggests that in many cases, they ignore each other. Behind the scenes though, there is a complex food web and there are obviously exceptions.

Jo has footage of a clan of badgers raiding a nest of western whip snakes, and foxes rooting up earthworms. When she looks up to the sky, at times she observes crows attacking buzzards in a proactive form of self-defence. Her two cats employ all their senses, wits and abilities to actively seek out and terrorise small rodents. In the early days, she was desperate to connect more personally with the wildlife there, now she is more conscious of the role of the human as predator: "We do not wish to desensitise wild creatures to our presence and by doing so, increase their vulnerability as they leave the relative safety of our place. Whilst we continue to wander in the woods, we too are learning the gentle art of avoidance and turning more to the trail camera as our eyes and ears."

Jo has established areas of wilderness (aka 'Zone 5') in the woods, where she never treads: "We view this as a precious ecosystem and safe space where creatures can escape the perils of interaction with humans." This has resulted in wildlife-rich **edges** and she is gradually modifying her maintenance strategy and the associated 'work' to minimise the interventions near those edges.

The nature of work

I don't think Bill Mollison was particular about choosing the word *work* in the title of this principle. It could easily have been 'to live with', 'be with' or 'care for'. But he chose 'work' and it seems to fit. From a permaculture perspective, work is a physical or mental effort that we have to do because a system doesn't do that work itself, and at the other end, pollution is a resource that a system is not utilising. So at one end of a system we need to put energy and time in to keep the system going and at the other end the unused resources may become wasted physical outputs. There could also be a mental waste as the toxins of frustration, resentment, and conflict with other priorities manifest. Typically in our modern societies we find lots of work and lots of pollution. I have often referred to this work and pollution exploration as ***work is pollution.*** I even thought Mollison was responsible for this phrase until I started researching for this book and I couldn't find it anywhere. Turns out that was just in my head. It sounds like one of Bill's zen sayings but it wasn't. I may have picked it up from Patrick Whitefield's book *Permaculture in a Nutshell*[10] where he refers to ***work and pollution.***

Work = any need not met by the system
And Pollution = any output not used by the system

Patrick Whitefield[11]

I typically find explorations to minimise work appearing in the maintenance end of a design process. We have designed a system that looks like minimal work but actually upon implementation it is more work than we bargained for to keep the system going. We know that if we can tweak it then hopefully our work load of maintaining the system can be reduced. There is potential here to overlap with the principle of ***least change.*** If we can just find that one perfect tweak and intervention that significantly lowers our maintenance workload, then bingo.

Minimising work

Is there anything that can help us to minimise our work? The most powerful of the natural forces and one that is always available, is gravity, and we can use that to automatically carry water around our site. The design of our water storage systems should use the principles of work with nature and ***relative location*** by storing the water at the highest possible location on our site so that it can be held back and fed by gravity to where it is needed. There are various strategies that can help us here, such as swales, dams and overflow pipes, all carefully positioned.

When we allow animals to do their thing and be natural then they might be there alongside us to help, after all ***everything gardens***. Maybe it's the worms turning and processing the soil, or birds eating bugs. Some permaculture designs can involve 'chicken tractors' or hens and pigs eating up vegetable kitchen scraps. And there are plenty of microbes and insects that help us to create compost. We may still need to turn it and check it but they are doing the majority of the work. As for plants, well they do a ton of work in catching sunlight which in turn can become food, fuel and fibre. We can utilise plant varieties that sow their own seeds or bring nutrients up from deep in the soil.

How do we feel about computers, robots and artificial intelligence? Many of us are using them to minimise our workload, being more ***efficient***, whether that be our smartphone for calculations, searching on the internet, storing photos, using online maps or automating information on spreadsheets. We can set up systems that look like minimal work, but do we really want to not work? After all, humans love work. We love being busy and occupied. Not to the point of overwhelm but definitely engaged and inspired to take up the challenge. It's great for our self-esteem, creativity and the desire to be productive. We love it when we can ***cooperate*** and work alongside others that motivate us and work with our own nature. Like many small farms we take on volunteers who love to get involved, meet up with friends, share food together and form a community.

We need to be careful about our expectations when it comes to work. For example, forest gardens look like minimal work and they have their moments when you casually gather a salad and a bit of fruit for lunch. But then we can be overwhelmed when all that fruit is ready to harvest and it needs processing, juicing, jamming, fermenting, pickling, drying and storing. Oh and possibly organising volunteers to help. There's also pruning, a bit of chop and drop, maybe mulching and maybe marketing and selling your excess produce. The scale of the challenge needs to fit our capacity.

In southwest France, Jo has chosen a laissez-faire approach to her food forest: "We view it as a resource to be shared with the creatures who were here before us, who live in ignorance of human land boundaries and ownership entitlements." She tries to minimise her interventions in sensitive areas, but is not idle. "We grow food to feed ourselves and we share surplus with friends. We harvest enough for our needs and preserve some of our harvest in the form of stored nuts, pickles, wine, and cider. We have created large wildflower borders, which we cut annually

Jo's fairshares heap © Jo Holleran

over a period of several months." Fruits are graded and those that are not suitable for her own consumption, such as windfall apples with heavy damage, are destined for the 'fairshares' compost heap, which is located at a busy junction of the 'wildlife superhighway' in the ravine. From here, she gains the ***yield*** of hundreds of trail-cam videos each year: what a great ***observational*** tool. "Often, as I walk quietly around our land, I hear the hustle and bustle of the hidden world in the unmanaged tangle of 'Zone 5' and I wonder what life is like for the creatures there, and if I am being watched."

"As my time and energy are both limited and because there is so much that I still want to achieve with whatever time I have left on Earth, I must use these limited resources wisely and efficiently. Nature has infinitely more wisdom and experience than I can even imagine and so it makes perfect sense to allow nature to set the direction, the momentum, and the rhythms of my work. I see myself walking alongside nature, using my mind and my body to the best of my ability, to perform appropriate interventions for as long as I am able. In return, I will take a yield which is sufficient to sustain myself and the other members of my household and do my best to limit the negative impacts of my modern-day existence."

Human nature

As humans we have patterns of thinking and feeling in common that we can regard as the nature of being human. We have a tendency in our cultures to think that those things are all quite different from the nature of other animals. We believe that we have added logic, ethics and reasoning to our range of tools but we also have the arrogance to think these things are uniquely human. We have pushed away, downgraded and at times even distrust our instincts and intuition, for these are the traits that we see as truly animalistic. The stepping stone towards going back to nature is to listen to your own intuitive voice, your gut reactions and the knowledge in your body. It can be a quiet voice so give it time and a space to be heard and exercised. We need to remember that we are part of nature and not separated from it.

As permaculture designers, we need to include these intuitive and instinctive aspects in our designs of logical approaches, planning of systems and reasonable decisions. We may be working on a design which doesn't involve land at all, which perhaps focuses on ***people care***, be that ourselves or our community. Involving our intuition and instincts in those designs might manifest as uncertainty, curiosity and trusting that quiet voice. The nature of intuitive and instinctive thought is that it is very rapid, coming from the depths of the unconscious. It doesn't appear as rational and it certainly doesn't appear to have been through a process of analysis.[12] That is because those quiet thoughts are from prior experiences where we have formed an opinion or judgement to support our own future survival. When I observe my own thoughts they can appear twice, first the quiet intuition and then a repetition, louder as it comes into my awareness and I give it attention in my conscious mind. Sometimes the need to act on these initial rapid thoughts is urgent, such as braking suddenly whilst driving to prevent a crash, the mind making the body move and bypassing any conscious choice. In other cases we have time to allow the thoughts to be analysed as there may be a better and more reasonable response.

Sometimes the instinctive thought might be plain wrong but convincing. Take care not to believe all of your own thoughts. We watch the sun apparently move across the sky and our hunch would be that it is the sun moving. It was only careful calculations and analysis that revealed that it is the Earth that moves. Intuition is great for looking out for ***patterns***, the moments when our bodies know we have been here before and know what to do next.

There are arguments to suggest that human nature is full of selfishness and evil. I prefer the theory from the non-violent communication founder Marshall Rosenberg that "behind every action, however ineffective, tragic, violent, or abhorrent to us, is an attempt to meet a need".[13] With skill, training and awareness we can find alternative positive strategies to fulfil those needs. It is not in our nature to destroy the planet we find ourselves on. We can be stewards and custodians of our environment and apply ***earth care***, we can give back and enable the regeneration of this biodiverse planet. We are, though, mainly due to the powers over us, such as governments and corporations, desperately meeting our needs for security, warmth, movement, nutrition etc. through actions that have terrible consequences for the planet. We need to include the ***small and slow*** principle in all of our actions to ensure that our footprint on this planet is a positive one.

Another social aspect of working with nature could include understanding other cultures and societies, of which there are many. We may encounter this when travelling abroad or when working in a sector different to our own. Ways of being and cultural norms will be different, which can lead to misunderstandings and even accidental insults. Trying to navigate a school or council when running a community project can be a minefield. Finding the right person within that organisation who understands their systems and can help you work with the nature of that group can be the best strategy.

Let's conclude this chapter with more from Jo Holleran as she shares her approach to the land. "Late March is a busy time of year in the potager. Every dry day is spent refreshing the beds with compost, sowing seeds, and applying mulch in preparation for a summer bounty. Most beds host a range of different fruits and vegetables, as we weave a complex polyculture tapestry that differs from year to year. A labelling system is invaluable, to ensure that we recognise the seedlings as they break through the surface and to ensure that we nurture their growth appropriately. Fortunately, we maintain multiple back-ups in the form of diary notes, sketches, and photos. One fresh March morning, I stepped out into the birdsong to survey and take pride in our achievements from the previous day. I was stunned to find that every carefully placed plant label in the potager had been lifted and strewn around. On closer inspection it was evident that most of the labels had journeyed from their original placement to another bed. I'd love to have seen the culprits in action! We set the trail camera up to discover who or what was behind this naughtiness. After a few nights of filming, we were delighted to watch two mischievous fox cubs treating the potager as their evening playground. A pattern of us resetting the labels and the cubs returning to uproot them continued for around two weeks, until they either established a more stimulating game or turned their attention to more grown-up matters. The game has been repeated every year since and so we can only think that it has been communicated from one generation to another. We feel as if we have formed a connection with this family, albeit via the trail camera as an intermediary, and without ever knowingly being in the same place at the same time. The practice of the gentle art of avoidance."

Yield Unlimited

Written in collaboration with Jo Barker and Graham Bell

The various yield principles

The principle ***obtain a yield*** was coined by David Holmgren.[1] It can be simply translated as any effort we make should aim to gain a yield. It follows on from Bill Mollison in *A Designers' Manual* coining the principle that the ***yield of a system is theoretically unlimited*** or at least only ***limited by the information and imagination of the designer.***[2] I prefer Bill's version of this principle as it clearly pushes us to explore our creativity, continuously tweaking and improving our designs to keep adding yields to our systems, such as adding in that extra grapevine in that vertical space now available. Rosemary Morrow's angle was to ***minimise maintenance and energy inputs to achieve maximum yields,***[3] Heather Jo Flores said that ***the designer limits the yield,***[4] and Toby Hemenway tells us to ***get a yield.***[5]

Why did Bill say that the yield is *theoretically* unlimited? Basically because, and David Holmgren agrees, that it is a contradiction of natural laws which state that energy is limited and can neither be created nor destroyed, and yet here we have a proposal to create something unlimited and potentially infinite. This makes it lean more towards being an attitudinal principle. We can define our imagination as infinite so why don't we just simply say that ***yields are unlimited.*** There is always scope to exercise our creativity using techniques such as lateral thinking. And even in our finite lives there is death and microbes that will turn the old and decayed into compost to support new life. As change is constant, there could always be the opportunity for a new perspective.

Jerusalem artichokes © Wilf Richards

My preference with this principle would be to rename it ***yield unlimited,*** or as Colleen Stevenson put it ***the harvest is unlimited,***[6] with the intention that we can just keep increasing our yields, hence the title of this chapter. This would link it more clearly with our permaculture approach of seeing that ***every element can provide multiple functions*** and our ongoing tweaks and maintenance of systems, pushing towards increased connections and beneficial relationships. For example, Jerusalem artichoke plants provide us with tubers to eat over winter, flowers in autumn and the long stems can be cut and dried easily in the winter for kindling. What else could that plant yield? Maybe farting competitions and knowledge of inulin and gut health. Maybe, believe it or not, a great recipe for créme brûlèe made using Jerusalem artichokes![7]

Yield as a goal?

Some permaculture designers have questioned the notion of yield being included as a principle, as really it should simply be seen as a goal. We can see a similar situation with ***functions*** too as they can easily emerge from aims and goals as well. Obtaining a yield should be one of our key goals in the first place by asking: What do I want? What do I need? Our goals could be food, shelter, profit or more abstract yields such as happiness or learning. Every design challenge will be approached differently by each of us, and so collectively, if we were to gather together all of our possible yield-based goals for a design, then maybe they too would be unlimited. By focusing on an *unlimited yield* as a constant tweak and improvement to our designs to ensure ever increasing yields, this switches this goal-based statement of *obtain a yield* more towards being a principle as an ongoing guide to encourage our creativity.

We can also see the intention to maximise yield in the behaviour of various animals and plants. For example, plants move their leaves during daylight hours so that they are facing the sun to maximise photosynthesis. Plenty of animals will ***catch and store*** their harvests when they are abundant and stash their hoards for a rainy day. We can also see it in the general intentions to either ***cooperate*** or compete with other species for the purposes of survival. The intention to increase yield is everywhere.

Are you sure there was no yield?

So what happens if, for example, in a food crop design we end up with no crops? A worst case scenario of extreme flood or drought that killed everything. Well we aimed for a yield of food but did not gain the yield we were looking for. In these situations we need to consider other yields. Was the experience the yield? Or learning about what not to do? Or the collaboration, joy, wellbeing and connection with the friend we tried to grow veg with? Having some sense of positive thinking, optimism, evaluation of lessons learnt and determination to try again with a different approach are key in such failings – keeping a gratitude log to note all of the things we can appreciate and to remind us that there is always something we have gained. This will provide us with the motivation to try again and also to acknowledge that a different approach may be required next time. After all, if the sun is still coming up every morning, you are still alive and the internet is still working, then really what is there to complain about? Maybe we need to consider the ***problem is the solution*** principle when no yield is gained. Maybe it was a pest that destroyed your crop or a rabbit that came in and ate all the carrots. Maybe there was a lack of predators to keep the pest down. Maybe someone will like to eat the rabbits. When the squirrels have eaten all of our hazelnuts, I can see the squirrels as a potential yield instead.

How to decrease your yields!

And why would you want to decrease your yield? You wouldn't, except maybe putting the brakes on the apparent infinite yield from social media. Our modern economic systems are bit by bit decreasing yields across the globe, despite claiming the exact opposite. Whether that be through soil degradation, land contamination, loss of biodiversityor the poisoning of our planet. Soil is being lost from modern intensive agricultural systems through over-use of the plough, periods of bare soil, over-use and dependence on fossil fuel derived chemicals and soil erosion. All in the name of maximising yields whilst actually destroying the very foundation of yield, the Earth itself. Plenty of data shows that small-scale gardens and smallholdings are more productive per acre than intensive agricultural systems. One of the reasons for that is the steady yielding throughout the year as compared to a single monoculture all ready at the same time from one massive field. Should the resources not be available to harvest, store, sell or process that singular yield then it can end up being left to rot in the field, a situation far too common after Brexit in the UK resulted in a lack of seasonal farm hands. The drive towards ever increasing economic growth is truly unsustainable and based on constantly exploiting the Earth to increase supposed yields. That is one big 'yield' that we do need to stop and shift towards sustainable regenerative yields instead.

Yield as net gain

There can be situations where we want to focus on the net yield of a system. This is one way in which we define what a yield is. We can see yield as the gross gain but we could also take into account the costs and other overheads and subtract these from our gross gains and thus focus on the net yield. For example, your hens may produce eggs. As they grow older the egg count will start to drop but the work input and the cost of the feed input may stay the same. This may well mean that your eggs start to become very expensive. Is it worth carrying on? Is it time to replace the older hens for younger ones? Is this a commercial operation or are they family pets? A review of an existing system to ensure that there is a net yield is important for motivation and net energy gain. How we define the yield will depend on the context.

Let's consider soil fertility. We may be adding fertility to the soil in the form of composts and manures but we may also be extracting more goodness from the soil than we are adding, through the growing of crops in that field. We need to consider the net gains of fertility to the soil. Ideally in this example we would put in place the conditions to enable ongoing building of soil fertility. We could apply the same thinking to our cash flow. It could be amazing to have an income of a million pounds a year but not if your overheads are two million pounds a year. You are making a loss and it is time for a rethink and a new design, maybe one that considers other principles first, such as ***limiting factors*** to increase yields and protecting what yields we already gain through ***catch and store energy***. We may also consider lowering costs and overheads, or in a different context reduce the work input whilst retaining the same yield, thus shifting towards ***energy efficiency***. As an example, this might include adding more perennial plants and self-seeding plants into your garden that provide a harvest for less work input, compared to the seasonal sowing and planting of annual plants.

To help ensure we have a net gain we need to measure our yields in the first place, starting with measuring baseline data and then measuring the difference so we can have some sense of what the gain has been. Find a way in your designs to log data and keep a record of the numbers. In our veg box scheme at Abundant Earth, we weigh out portions of veg for each customer and know how many customers there are each week and so from this we can calculate total veg production from the garden.

Yields before we even start

We can also obtain a yield before even designing or implementing. Many situations can provide a yield at the ***observation*** stage. Just having a look at the bare field you intend to grow food crops in can reveal the abundance of wild foods and medicinal plants already there. Or maybe the yield is simply learning about the history of the land, its previous occupants, the uses of it and the changes in appearance.

Yield, more yield and even more yield

One of the common intentions for permaculture designers is the creation of polycultures, by which we mean the opposite of monoculture. This is normally associated with just agricultural activities but could equally apply to many other systems, including social settings. Monoculture in agricultural terms is normally just focused on the two dimensions of width and depth, focusing on big fields of one crop and yields per acre. Permaculture goes way beyond this to consider at least five dimensions that includes the various heights of different crops, the complex interactions between the differing crops and making the most of changes through time, in particular the seasons. Obvious examples of this include forest

The record-breaking super salad © Jo Barker

gardens and vertical growing of food crops. There is a clear overlap here between ***diversity*** and increasing yield. One of the reasons that increasing diversity of our crops increases yield is that the variety of plants have different needs. We can, if we get the design right, decrease the likelihood of plants competing with each other. One may want the light, whilst another thrives in the shade. Meanwhile a variety of root depths will mean differing access to water and minerals.

Jo Barker has pushed yield to the limit with her creativity. First of all she developed Dave Jacke's list of seven functions and yields of forest gardening (food, fuel, fun, (f)pharmacy, furniture, fodder and fibre) and took it to over seventy, and still managed to keep most of them beginning with the letter f. Then when Jo was asked to design an orchard during the Covid pandemic, she took the opportunity to list the many yields of apples. Jo created a mind map and animation to show all of the possibilities.[8]

Jo also holds the unofficial record for the most ingredients in a UK salad. She has been foraging and making edible gardens for decades. At first she was amazed to find 20-30 edible things for a salad in most gardens everyday of the year.

Then it was 50, then 80, and then knowledge seemed to jump in leaps to 120, then 180, 240 and finally 325 was the last record-breaking salad in May 2019, foraged in the Greenhouse Food Forest in Ramsgate. It included different parts of the same plants such as flowers, leaves, tips, fruits and roots. She stopped asking "Can I eat this?" and instead asked "How do I eat this?". These super salads have increased nutrition and medicine that is not found in shop-bought food. They also add diversity to a garden and inspire others. It took an hour and a half to harvest and record the ingredients of this beautiful salad.

An ethical yield

Let's end on a note of caution though, as this is one principle that the capitalist world loves, even if they tell lies about their yields. They love nothing more than increasing yields but at what cost, and here is an example of when a principle needs to sit alongside and be embedded with the permaculture ***ethics***. For if we do not also consider ***earth care***, ***people care*** and ***fair shares*** alongside our intentions to ever increase our yields then we are destined for a dead planet and failure of the human species. Our intentions are clear: increase the yields but not to the detriment of the earth or all those that dwell on her. We can still increase our yields carefully and reach that point where we can ***share our surplus*** as the more yields we have, the more we have to share.

How I Used the Principles to Write This Book

I, of course, used many of the permaculture principles to direct and manage the writing of this book over a period of time spanning from about March 2022 through to November 2024. I also created a design, which is not included in this book, to enable me to plan and reflect upon the journey.

Catch and store everything

The first obvious principle I used right from day one was ***catch and store energy***. Every time I had a useful thought, idea, relevant conversation, found a useful website or poster or handout then I would log it, bookmark it and add it into my notes document. Every relevant email was noted, recorded or downloaded. I have been using a combination of Evernote, Trello and Google Drive for years now and find them remarkable tools for storing and organising information. In the case of this book, I focused on only using google drive to develop various documents and folders, with a separate document for each chapter.

Get the patterns in place

This meant that I had to have a way to organise and manage that information and that is where ***pattern to details*** came in handy. I developed folders and folders within folders to store the various chapters, documents, photos and files. This enabled ease of access, quick retrieval and the ability to relocate files if necessary. I identified key headers and subheaders to work with this principle and of course those evolved over time. In the early stages I would have to recheck the pattern and detail level to check it all still made sense. There was also an aspect of ***pattern to detail*** in each chapter as I identified key questions and criteria I wanted to answer and fulfil. That gave me structure and frameworks to hang items off and ensured there was a consistency across the chapters.

I obviously had to have ***functions and elements*** for the book. The functions started with broad goals, such as why write this book and what is the need for it. Out of that grew clearer detail-level functions such as identifying definitions, finding origins, exploring applications, critical thinking, collaborations and the illustrations required. These in turn became criteria for each chapter so I could check that broadly speaking I am fulfilling every function in every chapter.

The elements are of course the chapters themselves but also include the resources and collaborators. I developed a colour-coded spreadsheet with a scoring system to map this out which enabled me to see at a quick glance which chapter was in the lead and which one needed the most work.

This enabled an ***energy efficiency*** in the project, especially the intention to ***only handle it once.*** There has never been any rough paper based notes. Everything has gone straight onto Google Drive. In most cases information was stored in a rough notes Google document first, which was organised to include a section for each chapter. Then when I worked on a particular chapter I had those notes to work from.

There was a potential for each chapter to sit isolated, unconnected to its neighbouring chapters. That would never do. And so ***integration rather than segregation*** was used to go back to the pattern level and ensure crossovers between chapters were fulfilled and highlighted. I developed a principles map to show the key principles used and mentioned in each chapter and which ones still needed to be mentioned.

Start small and slow then scale up

The whole project ***started small and slow*** by simply being handouts for my new format permaculture design course (PDC). During lockdowns of 2020 and 2021, I carried out a review of my PDC teaching design and decided to reformat it into a modular system. That resulted in having to change some of my handouts, especially the ones about the principles. Beforehand, I had a principles handout for each day of the standard 12-day PDC, each one covering three or four principles. In the new modular format I needed individual handouts for each principle because each principle could feature in more than one module. The new modular format gave the option for participants to attend as many modules as they wanted, depending if they just wanted a taster for one day or to attend many modules to gain their certificate. I soon realised that I was accidentally creating a booklet on the principles and potentially a book. This was a gradual approach to begin with, updating one handout at a time but as the book developed I realised that ***small and slow*** was not going to work. I could no longer do this at the fringes of my work hours.

The project grew more complex and, in mid 2022, I started to commit a few hours a week of work time to the task. By early 2023, this grew to Saturdays as well, and by mid May 2023 I had to cut back on voluntary activities and some paid work to increase work on the book to three afternoons a week and all day Saturdays where possible. Every time these jumps in commitment took place I was primarily analysing and ***identifying the limiting factor*** of time, which prevented me from taking the project forward at the speed I wanted to go. I was ***adapting*** my approach and ***responding to the feedback*** to myself on the book's progression. I was also actively and gradually minimising my interaction with anything unconnected to the book. In that way I was ***maximising my edge*** with the book and ***minimising my edge*** with anything that could be a distraction. By October 2023, the work had grown to several hours every day. This was definitely a sensation of galloping along and certainly not going slowly anymore. I had instead to keep an eye on my wellbeing, an aspect of ***people care***, and put ***limiting factors*** in place such as not working before or after certain hours, having regular breaks and stretching my body, thus ***self-regulating***, avoiding burnout and maintaining my ***resilience*** for the project.

By mid November 2023, I hit a wall. The original deadline for the book was 15th November and I was close but knew that I needed at least another month and with Christmas around the corner I thought two months was more realistic. I decided to just simply stop for a week and apply some ***people care*** and ***self-regulation***. I had a cough and a bad back and needed to rest and recuperate. I was quite actively ***inefficient*** and applied some ***least change*** by not doing any work on the book for a good week. At this time of year it turns colder too, the nights draw in and there is a drop in our electricity production from our solar panels. So needing to sort firewood, working within the ***limiting factor*** of lower electricity availability and a certain degree of seasonal ***adaptation*** as well all enabled me to maintain my ***resilience*** for the project. When I restarted I devised an 'easy way back in action plan' which focused on the ***small and slow*** bits that were easier to do, such as writing this paragraph. I rejigged my time commitment to the book, lowering it significantly, and focused on chapters that were nearly finished rather than ones that needed significant further work. And finally I called upon some key collaborators to help me get some of the difficult chapters sorted.

Destroying the status quo of individuality

Cooperation has been a key aspect of this book from day one. I knew I could not be solely responsible for writing a book about such an important aspect of permaculture. I am very familiar with working in groups and collaborating as part of various roles that I have, including Abundant Earth coop member, Green Durham[1] director, permaculture teacher, diploma tutor, designer, coordinating volunteers and being in the band Queen's Screech. I knew that all of the permaculture designers and teachers had something to contribute towards this book – I wanted to ***integrate*** them into the book. I have been very involved in the development of the diploma system in Britain since about 2009. Inevitably this means that my go-to collaborators have been my fellow diploma tutors and diploma apprentices. We see a lot of designs, we assess them, give feedback and nurture the next generation of permaculture designers – they do amazing work. We are frequently blown away by the incredible projects undertaken and the commitment to personal, community and ecological resilience and sustainability. We frequently see a variety of interpretations about many aspects of permaculture including the principles of course. I am truly grateful to everyone who has been involved in this book whether it was just attending a workshop I ran, adding a little comment, reading a chapter, suggesting edits, researching with me, helping to edit a chapter or having conversations to explore the depths of a principle. Thank you, all of you.

One of the principles that has been a struggle for me was ***diversity***. On one hand the diversity of voices in this book is high with so many people being involved. But by focusing on the immediate colleagues around me and reaching out to other tutors and teachers through existing networks has meant that this book is predominantly written from a British/Northern European perspective. Once I had around 40 collaborators, I realised I was hitting a ***limiting factor*** on my time and capacity to manage communications with a lot of people. I applied a positive bias to increase involvement by female collaborators wherever I could. By the end of the book, I had active collaboration with just over 40 people, with over 25 of those being female. I would be interested to know how this book is perceived in other continents and by other cultures around the world and I hope it is not rejected on the grounds that there were not more diverse voices included. I have deliberately left open the potential for debate and discussion about the principles. I encourage you all to make them yours: no one owns them.

The multiple functions and efficiency of researching and writing

Looking up data, listening to the radio, reading websites, organising Zoom calls – there has been a lot to do, and there are times when I have had the opportunity to ***stack functions***. This could be as simple as washing up whilst listening to that relevant radio programme. Then there have been the times when researching for one chapter has also helped another chapter at the same time. There is a certain degree of ***energy efficiency*** around this process, which is also expressed through the deliberate focus on one chapter at a time. Typically for each work session I would work on the least developed chapter and get it up to speed by processing the comments made by the collaborators and ***integrating*** my rough notes into that chapter. I would then send it back to the collaborators and move onto the next chapter with the worst score. Each chapter also had ***patterns and details*** by providing bigger-picture perspectives and gradually working towards examples or lists. In the middle of October 2023 I also had another opportunity for ***multiple functions*** when I ran a course on the principles. This gave me an opportunity to review the book so far and create flip charts as visual displays for the course, which I also photographed and have included in relevant chapters.

Can I make it any better?

When we request reflection on a permaculture design we are asking what went well with the tools we used, what did we learn from the tools we used and what could have made it even better? One of those tools we use in our designs is the principles. Were there any principles that I could have used in writing this book that would have made it even better? Inevitably I have a bias and blind spots so they are hard for me to see. It seems to have gone quite well! To do this task effectively we need to have a tool that enables us to think outside of the box, such as just simply going through lists of principles and asking which ones have I used actively (as written about above) and which ones have I missed or could have used differently or even more.

I checked through my principles map to explore this and to actively see which principles I could mention or use further. I could have talked about proofreading as a form of ***producing no waste*** and how the involvement of collaborators was encouraging the contributors to garden the book into existence, as ***everything gardens***. I consider that I have not given any thought at this point in time to ***renewable resources***, after all this book will involve the use of lots of paper and ink to print and has certainly involved all of the electricity used at home to power my laptop (all off-grid solar power so minimal issue there) and all the power that I have used to keep everything stored by using Google Drive (Google states they've been carbon neutral since 2007, so maybe that is OK).

I seem to have not utilised ***least change***, apart from the pause in November 2023. Maybe if I had stopped at just the leaflets for my PDC and self-published that then I could justify least change. This book has certainly not been minimal work. I have not mentioned ***observe and interact***, although I obviously have done it as part of the process of research and writing. And ***relative location*** gets no mention in my exploration of the principles used either. Maybe that is because I decided from an early stage to put the principles in alphabetical order rather than trying to work out an elaborate order for them. The idea of crafting an order to the principles seemed like hard work: impossible and futile as they all connect somewhere. I also did not mention ***work with nature***, although you could say that the entire body of work is about supporting the reader to work with nature. And finally the ***problem is the solution*** didn't get used either. In any design and application we are unlikely to utilise every principle, and so just as I have reflected upon here, we can see that certain principles have been key, some useful and some barely used at all.

A retrospective exploration of emergence

At the 2023 Permaculture Convergence on the outskirts of London, I had a conversation with Looby Macnamara about utilising her book *Cultural Emergence*[2] and the principles it contains to support the process of writing this book. I had to admit at that point in time, I had not given them much attention apart from reading about them and seeing some designs where they had been used. I had also been very inspired by and impressed with Looby's freedom with the principles: "You can create new principles for anything you do… cooking, tree climbing, song writing… it can help us reduce mistakes, for ourselves and others and decrease the need for trial and error… new principles contribute to collective wisdom that we can share with others. They become easy reminders that provide us with shortcuts, improving our effectiveness."

I decided to see which ones I had accidentally used up to this point and which ones had not been used and if they could help in any way. Bear in mind that the deadline for the final draft of the book was in one month's time. Looby shared with me the latest principle to be added to that set: ***make connection first.***

There are seven sets of cultural emergence principles. The first set is called 'principles for growing inner wisdom'. They include ***your body knows***, ***honour the wisdom within*** and ***tend to your personal culture.*** These ones do not massively speak to me in this context. I have been largely focused outwards with this project. There are times when I have listened to my body of course, such as stopping and resting where required.

The second and third set both felt strongly activated already. The second set is called 'principles for creating fields of encouragement'. They include ***be in courage***, ***give encouragement***, ***receive encouragement*** and ***use the intelligence of co-operating hearts.*** It has been a big leap of faith to write a book, it's my first one. And I have been encouraging collaborations and contributions right from day one, and receiving plenty of encouragement from my collaborators. I have a lot of gratitude for those that have contributed. The third set of cultural emergence principles, 'principles for strengthening cooperatio', also seemed obvious. They include ***anyone can raise the vibration***, ***emergence happens in relationship***, ***many minds are better than one*** and ***weave unity.*** C ooperation with my collaborators has been key. Right from the start I knew that I could not do this alone. I have enjoyed the abundance of emails, comments, Zoom calls and conversations. There have been plenty of times when someone else's contribution has been key, adding the next layer of excitement and I like to think the whole process has contributed to the collective community of permaculture tutors and teachers that I have connected with.

Set number four is called 'principles for aligning'. They include ***respond to life***, ***be attentive to timing***, ***be attentive to shifts, openings and opportunities*** and ***synchronise with natural patterns and cycles.*** Timings have been key, after all this book has an imposed deadline for the final draft. There have been plenty of moments of openings and opportunities to connect with other people, including David Holmgren. And I have utilised the natural pattern of working harder as we approach a deadline as I have already described in connection to starting ***small and slow.*** The fifth set seems closer to the work so far. They are called the 'principles for inviting flow'. They include ***trust the process***, ***step into the unknown***, ***use emergence to support emergence***, ***plan then flow*** and ***leave space for emergence.*** To start the book I created a design, a plan that led to flow. I have certainly stepped into the unknown on a regular basis, either in conversation with someone or in the development of a chapter and not knowing where it is going. And plenty has emerged over the time of writing and I have encouraged it.

Set number six is called 'principles for being proactive'. They include ***move the tools***, ***presence in the process***, ***make progress visible***, ***work to completion*** and ***beauty in completion.*** My work has been entirely visible being stored in Google Drive. Many of the collaborators will have seen the progress and the effort to work towards the final goal. As described further above I have had a spreadsheet to help me identify the criteria for each chapter and that has enabled me to see both beauty and steady progress towards the end. And as I head towards the finish line the principle of ***work to completion*** keeps poking me.

The last set, number seven is called 'principles for emerging potential', they include ***come into the light***, ***discover abundance***, ***embrace potential*** and ***allow for the possibility of the seemingly impossible***. These all strike a ***yield unlimited*** chord with myself. I have been repeatedly impressed with the abundance of support and the abundance of material to inspire aspects of this book. The potential of all the collaborators has been enormous and the book itself seemed impossible to begin with. The yields have been great, deeper connections with people and lots of learning and new insights into the principles. So from a retrospective consideration it looks like I have utilised most of the cultural emergence principles at some level.

The journey hasn't finished yet

I like to think that I have ***obtained a yield*** but that there is more, as of course the ***yield is unlimited***. I hope this book inspires further writing about the principles especially amongst cultures that I did not reach. And of course I love and ***accept feedback***: you are bound to have some. Let's have a conversation about it and keep the discussion about the principles growing and developing. You can email me at wilf.abundantearth@gmail.com or visit https://linktr.ee/wilf.richards.

Endnotes

Introduction, pages xi-xii

1 David Holmgren: *Permaculture: Principles and Pathways Beyond Sustainability,* Melliodora, 2002
2 https://thelandmagazine.org.uk/articles/permaculture-big-rock-candy-mountain
3 Patrick Whitefield: *The Earth Care Manual: A Permaculture Handbook for Britain and Other Temperate Climates,* Permanent Publications, 2004, p.13
4 Robert Anton Wilson interview www.youtube.com/watch?v=QvF8ekcyFvM

Delivering the Principles, pages 1-7

1 https://en.wiktionary.org/wiki/principle
2 https://integralchurch.wordpress.com/2012/07/10/15-great-principles-shared-by-all-religions
3 www.expii.com/t/scientific-principle-definition-examples-10310
4 Bill Mollison: *Permaculture: A Designers' Manual,* Tagari Publications, 1988
5 *Permaculture: Principles and Pathways Beyond Sustainability,* p.xxiv
6 *Permaculture: Principles and Pathways Beyond Sustainability,* p.xxv
7 BBC Radio 4, Analysis: Precedents or Principles? www.bbc.co.uk/sounds/play/b04p7ygh
8 BBC Radio 4, Crowdscience, what does a sustainable life look like? www.bbc.co.uk/sounds/play/w3ct4y4h
9 Article by Michael Hoag 14th Dec 2021 https://transformativeadventures.org/2021/12/14/why-i-cant-say-indigenous-people-invented-permaculture
10 Bill Mollison: *Permaculture: A Designers' Manual,* p.10
11 Eugene N. Anderson: *Ecologies of the Heart: Emotion, Belief and the Environment,* Oxford University Press, 1996
12 Permaculture Research Institute, article by Graham Bell, Bill Mollison: A Recognition of his Life www.permaculturenews.org/2016/09/27/a-world-class-hero
13 Bill Mollison: *Pamphlet VIII in the Permaculture Design Course Series,* Yankee Permaculture, 1981, p.73
14 Tyson Yunkaporta: *Sand Talk: How Indigenous Thinking Can Save the World,* Harperone, 2020
15 *Permaculture: Principles and Pathways Beyond Sustainability,* p.22
16 Fikret Berkes: *Sacred Ecology,* 3rd edition, Routledge, 2012
17 Bill Mollison: *Permaculture: A Designers' Manual,* p.12
18 Bill Mollison: *Pamphlet I in the Permaculture Design Course Series,* Yankee Permaculture, 1981, p.8
19 https://en.wikipedia.org/wiki/Regenerative_agriculture
20 www.rase.org.uk/news/the-principles-of-regenerative-agriculture
21 www.ifoam.bio/why-organic/shaping-agriculture/four-principles-organic
22 www.bioregional.com/one-planet-living
23 https://en.wikipedia.org/wiki/Janine_Benyus
24 www.learnbiomimicry.com/blog/biomimicry-lifes-principles
25 Permaculture Research Institute, Clarifying the components of permaculture, Damien Bohler, 2017 www.permaculturenews.org/2017/09/29/clarifying-components-permaculture
26 Rosemary Morrow: *Earth Restorer's Guide to Permaculture,* Melliodora Publishing, 2022, p.8

Using the Principles, pages 8-13

1 Rebuilding our compost system at Abundant Earth https://drive.google.com/file/d/1zIw7UgYOvdPcNFbuhPtRqGGzMtEWOd8A/view?usp=drive_link
2 Stephen Andrews Diploma portfolio https://thepollengardens.com
3 Looby Macnamara: *People & Permaculture,* Permanent Publications, 2012
4 https://media2-production.mightynetworks.com/asset/12582477/Permaculture_principles-Starhawk.pdf
5 https://cultural-emergence.com
6 www.patchoftheplanet.co.uk/diplomaportfolio
7 www.permaculturedesign.earth/designdeck
8 Herb and flower bed design using web of principles by Tomas Remiarz www.permaculture.org.uk/content/zone-1-herb-and-flower-bed-web-principles
9 Harvesting habits design using web of principles by Neil Kingsnorth www.permaculture.org.uk/content/harvesting-habits-animal-systems-design
10 Permaculture Association (Britain) design library www.permaculture.org.uk/diploma/designs

Adaptation, pages 16-22

1 David Holmgren: *Permaculture: Principles and Pathways Beyond Sustainability,* Melliodora, Australia, 2002
2 https://permacultureprinciples.com/permaculture-principles/_12

3 Ian Lillington: *The Holistic Life: Sustainability through Permaculture*, Axiom, 2007
4 Bill Mollison with Reny Mia Slay: *Introduction to Permaculture*, Tagari, 1991
5 www.darwinproject.ac.uk/people/about-darwin/six-things-darwin-never-said#quote1
6 https://en.wikipedia.org/wiki/Impermanence
7 Rupert Gethin, *The Foundations of Buddhism*, Oxford University Press. p.74
8 https://en.wikipedia.org/wiki/Adaptation
9 https://en.wikipedia.org/wiki/Peppered_moth_evolution
10 www.yellowstonepark.com/things-to-do/wildlife/wolf-reintroduction-changes-ecosystem
11 www.theguardian.com/environment/2008/jul/20/fishing.wildlife
12 https://royalsocietypublishing.org/doi/10.1098/rsos.161040
13 BBC Radio 4, The Path of Least Resistance www.bbc.co.uk/sounds/play/b0138361
14 BBC Radio 4, Naturebang, Dragon Lizards and the gender spectrum www.bbc.co.uk/sounds/play/m000q3ks
15 BBC Radio 4, Start the Week, Ancient Trees www.bbc.co.uk/sounds/play/m001l940
16 BBC Radio 4, In Our Time, homo erectus www.bbc.co.uk/programmes/m00168lg
17 BBC Radio 4, The Climate Question, Can animals evolve to deal with climate change? www.bbc.co.uk/sounds/play/w3ct3kjg
18 https://en.wikipedia.org/wiki/Jack_of_all_trades
19 https://martincrawford.substack.com/p/apple-varieties-for-the-future
20 www.theguardian.com/global/2016/jan/31/mysteries-of-the-rhubarb-triangle
21 Natural Flood Management in the Forest https://heartofenglandforest.org/news/natural-flood-management-forest

Catch and Store, pages 23-27

1 David Holmgren: *Permaculture: Principles and Pathways Beyond Sustainability*, Melliodora, 2002
2 Bill Mollison: *Permaculture, A Designers' Manual*, Tagari Publications, 1988, p.14
3 https://wordhistories.net/2016/09/20/make-hay
4 https://idiomation.wordpress.com/2011/05/04/save-for-a-rainy-day
5 https://wildlifeinformer.com/animals-that-store-food
6 https://arboriculture.wordpress.com/2016/05/04/the-eurasian-jay-and-acorns-a-symbiosis
7 www.1900s.org.uk/1900s-storing-root-veg.htm
8 Sandor Katz: *Wild Fermentation: The Flavor, Nutrition and Craft of Live-Culture Foods*, Chelsea Green Publishing, 2003
9 https://en.wikipedia.org/wiki/Cruachan_Power_Station
10 https://polarnightenergy.fi/sand-battery
11 https://crmpi.org
12 https://organisemyhouse.com/diy-memory-jar-printable

Cooperation, pages 28-35

1 Bill Mollison: *Permaculture, A Designers' Manual*, Tagari Publications, 1988, p.2
2 David Holmgren: *Permaculture: Principles and Pathways Beyond Sustainability*, Melliodora, 2002
3 David Holmgren: *Permaculture: Principles and Pathways Beyond Sustainability*, Melliodora, Australia, 2002, p.167
4 www.youtube.com/watch?v=MS-XKMN95dk
5 www.thinklikeatree.co.uk
6 https://cultural-emergence.com
7 https://smarterthancrows.wordpress.com/2018/08/05/the-other-permaculture-principles
8 BBC Radio 4, In Our Time, Parasitism www.bbc.co.uk/sounds/play/b08bb9cy
9 BBC Radio 4, The Infinite Monkey Cage, the wood wide web www.bbc.co.uk/sounds/play/m00193mg
10 https://en.wikipedia.org/wiki/Peter_Kropotkin
11 BBC Radio 4, Cautionary Tales with Tim Harford, The Deadly Airship Race www.bbc.co.uk/sounds/play/p0gy6hl8
12 https://en.wikipedia.org/wiki/Nonviolent_Communication
13 David Holmgren: *Permaculture: Principles and Pathways Beyond Sustainability*, Melliodora, 2002, p.56
14 www.abundantearth.coop
15 https://en.m.wikipedia.org/wiki/Rochdale_Society_of_Equitable_Pioneers
16 www.radicalroutes.org.uk
17 www.uk.coop
18 www.uk.coop/news/uks-democratic-economy-hits-ps879-billion-people-turn-co-operatives-and-mutuals
19 https://permacultureprinciples.com/permaculture-principles/_8
20 BBC Radio 4, Sideways, Best Feet Forward www.bbc.co.uk/sounds/play/m000ydlx
21 BBC Radio 4, The Forum, When Does Healthy Competition Become Destructive? www.bbc.co.uk/sounds/play/p03vz83d

Diversity, pages 36-43

1 www.ceh.ac.uk/why-do-soil-microbes-matter
2 https://en.wikipedia.org/wiki/Gut_microbiota
3 Bill Mollison and David Holmgren: *Permaculture One, A Perennial Agriculture for Human Settlements*, Tagari Publications, 1978
4 www.motherearthnews.com/homesteading-and-livestock/permaculture-design-part-1-zmazz84jazloeck
5 Bill Mollison with Reny Mia Slay: *Introduction to Permaculture*, Tagari, 1991
6 David Holmgren: *Permaculture: Principles and Pathways Beyond Sustainability*, Melliodora, 2002
7 www.languagecouncils.sg/goodenglish/resources/idioms/dont-put-all-your-eggs-in-one-basket
8 Patrick Whitefield: *The Earth Care Manual: A Permaculture Handbook for Britain and Other*

Temperate Climates, Permanent Publications, 2004
9 https://en.wikipedia.org/wiki/Functional_equivalence_(ecology)
10 www.unep.org/news-and-stories/story/how-chernobyl-has-become-unexpected-haven-wildlife
11 www.motherearthnews.com/organic-gardening/self-pollinating-apples
12 David Holmgren: *Permaculture: Principles and Pathways Beyond Sustainability*, Melliodora, 2002, p.214
13 www.youtube.com/watch?v=36CKsP9YQ1E
14 www.debonogroup.com/services/core-programs/six-thinking-hats/
15 https://hbr.org/2017/02/diversity-doesnt-stick-without-inclusion
16 www.ktshepherdpermaculture.com/store/p88/Thrivingprincipleszine.html
17 BBC Radio 4, What If Everyone Was Disabled? www.bbc.co.uk/sounds/play/m000kx1l
18 https://tim-spector.co.uk
19 www.weforum.org/agenda/2016/01/why-do-we-consume-only-a-tiny-fraction-of-the-world-s-edible-plants
20 www.sociocracyforall.org

Edge Optimised, pages 44-50

1 https://en.m.wikipedia.org/wiki/Ecotone
2 https://en.m.wikipedia.org/wiki/Alfred_Russel_Wallace
3 https://en.m.wikipedia.org/wiki/Edge_effects
4 https://en.wikipedia.org/wiki/Coral_reef
5 https://en.m.wikipedia.org/wiki/Wallace_Line
6 Bill Mollison and David Holmgren: *Permaculture One, A Perennial Agriculture for Human Settlements*, Tagari Publications, 1978, p.29
7 Bill Mollison with Reny Mia Slay: *Introduction to Permaculture*, Tagari, 1991, p.26
8 https://en.wikipedia.org/wiki/Chinampa
9 https://deepgreenpermaculture.com/permaculture/permaculture-design-principles/10-edge-effect/
10 Patrick Whitefield: *The Earth Care Manual: A Permaculture Handbook for Britain and Other Temperate Climates*, Permanent Publications, 2004, p.24
11 Rosemary Morrow: *Earth User's Guide to Permaculture*, 2nd edition: Permanent Publications, 2006
12 https://tobyhemenway.com/resources/ethics-and-principles
13 www.youtube.com/watch?v=MS-XKMN95dk
14 www.thinklikeatree.co.uk
15 David Holmgren: *Permaculture: Principles and Pathways Beyond Sustainability*, Melliodora, Australia, 2002, p.223
16 https://permacultureprinciples.com/permaculture-principles/_11
17 www.cafre.ac.uk/business-support/agriculture/environment/environment-technical-support/scrapes-and-bog-flush-dam-ponds-on-upland-sites-for-breeding-waders
18 www.nhm.ac.uk/discover/why-road-verges-are-important-wildlife-habitats.html
19 https://en.m.wikipedia.org/wiki/Riparian_zone
20 https://en.wikipedia.org/wiki/Crannog
21 www.livescience.com/animals/snails/green-banded-broodsac-the-brain-hijacking-parasite-that-creates-disco-zombie-snails
22 www.gardenersworld.com/plants/companion-planting-combinations
23 https://en.wikipedia.org/wiki/Flow_(psychology)
24 www.runnersworld.com/uk/training/motivation/a27718661/what-is-80-20-running

Energy Efficiency, pages 51-59

1 https://en.wikipedia.org/wiki/Laws_of_thermodynamics
2 Bill Mollison and David Holmgren: *Permaculture One, A Perennial Agriculture for Human Settlements*, Tagari Publications, 1978
3 Bill Mollison: *Permaculture: A Designers' Manual*, Tagari Publications, 1988, p.9
4 Bill Mollison with Reny Mia Slay: *Introduction to Permaculture*, Tagari, 1991
5 www.motherearthnews.com/homesteading-and-livestock/permaculture-design-part-1-zmazz84jazloeck/
6 www.permaculture.org.uk/practical-solutions/energy-efficiency
7 David Holmgren: *Permaculture: Principles and Pathways Beyond Sustainability*, Melliodora, 2002
8 https://en.wikipedia.org/wiki/Energy_efficiency_in_transport
9 BBC Radio 4, CrowdScience, human v machine www.bbc.co.uk/sounds/play/w3ct4y45
10 https://en.wikipedia.org/wiki/Ecological_efficiency
11 Original quote from Arthur Koestler, *The Ghost in the Machine*, Hutchinson, UK, 1967. Quote was used by Bill Mollison in *A Designer's Manual*
12 https://energy-efficient-products.ec.europa.eu/ecodesign-and-energy-label/product-list/fridges-and-freezers_en
13 BBC Radio 4, The Life Scientific, Brenda Boardman www.bbc.co.uk/sounds/play/m000zt8v
14 www.statista.com/statistics/322874/electricity-consumption-from-all-electricity-suppliers-in-the-united-kingdom
15 https://en.wikipedia.org/wiki/Embodied_energy
16 www.sustained.kitchen/latest/2021/5/26/which-is-more-sustainable-dried-beans-vs-canned-beans
17 https://media2-production.mightynetworks.com/asset/12582477/Permaculture_principles-Starhawk.pdf
18 https://en.wikipedia.org/wiki/Rebound_effect_(conservation)
19 https://en.wikipedia.org/wiki/Consumerism
20 https://solar.lowtechmagazine.com/2018/01/bedazzled-by-energy-efficiency
21 www.energysufficiency.org
22 https://rmi.org/our-work/areas-of-innovation/office-chief-scientist/10xe-factor-ten-engineering

23 Robert Pozen: *Extreme Productivity: Boost Your Results, Reduce Your Hours*, Harper Business, 2012
24 https://healthymindsacramento.com/2018/06/24/only-handle-it-once-an-organizational-game-changer-for-adhd
25 https://blogs.mtu.edu/improvement/2014/09/12/ohio-only-handle-it-once

Everything Gardens, pages 60-66

1 Bill Mollison: *Permaculture: A Designers' Manual*, Tagari Publications, 1988
2 www.permaculture.org.uk/principles/everything-gardens-or-modifies-its-environment
3 https://en.wikipedia.org/wiki/Gaia_hypothesis
4 https://en.wikipedia.org/wiki/Animism
5 Bayo Akomolafe www.youtube.com/watch?v=bBVAYzBteIo
6 www.gardenersworld.com/plants/10-companion-plants-to-grow
7 https://en.wikipedia.org/wiki/Slug
8 https://lesliehalleck.com/blog/everything-gardens-a-very-useful-permaculture-principle
9 https://en.wikipedia.org/wiki/Mole_(animal)

Functions and Elements, pages 67-74

1 Bill Mollison with Reny Mia Slay: *Introduction to Permaculture*, Tagari, 1991
2 Bill Mollison and David Holmgren: *Permaculture One, A Perennial Agriculture for Human Settlements*, Tagari Publications, 1978, p.7
3 www.motherearthnews.com/homesteading-and-livestock/permaculture-design-part-1-zmazz84jazloeck
4 Patrick Whitefield: *Permaculture in a Nutshell*, Permanent Publications, 1993
5 www.thinklikeatree.co.uk
6 David Holmgren: *Permaculture: Principles and Pathways Beyond Sustainability*, Melliodora, 2002, p.155
7 https://deepgreenpermaculture.com/permaculture/permaculture-design-principles/3-each-important-function-is-supported-by-many-elements
8 www.rootsnpermaculture.com
9 www.bmc.com/blogs/resiliency-vs-redundancy
10 www.theguardian.com/business/2022/dec/12/english-hospitals-emergency-plans-power-nhs-trusts
11 https://savoursoilpermaculture.com/wp-content/uploads/2021/10/the-principles-of-permaculture-bill-mollison-david-holmgren-.pdf
12 BBC Radio 4, Cautionary Tales with Tim Harford, La La Land: Galileo's Warning www.bbc.co.uk/sounds/play/p0gy6fs6
13 https://deepgreenpermaculture.com/permaculture/permaculture-design-principles/2-each-element-performs-many-functions
14 https://docs.google.com/presentation/d/0B-BjIk7BCT5zejZ1VlBud3VpVnZiaUw5Z3RuVTRPMWJVRlg4/edit?usp=drivesdk&ouid=112855547874266999616&resourcekey=0-28oHHLobY1vaLmWrCHZ6Lg&rtpof=true&sd=true
15 https://drive.google.com/file/d/1_G_my9OsUeB-b19gdOpszI4BThbcc_Qs/view?usp=drive_link
16 https://arnaudfache.wordpress.com
17 Aranya: *Permaculture Design: A Step-by-Step Guide*, Permanent Publications, 2012
18 Aranya: *Permaculture Design: A Step-by-Step Guide*, Permanent Publications, 2012

Integration Optimised, pages 75-80

1 Bill Mollison and David Holmgren: *Permaculture One, A Perennial Agriculture for Human Settlements*, Tagari Publications, 1978
2 Tomas Remiarz: *Forest Gardening in Practice; An Illustrated Practical Guide for Homes, Communities and Enterprises*, Permanent Publications, 2017, p.15
3 David Holmgren: *Permaculture: Principles and Pathways Beyond Sustainability*, Melliodora, 2002
4 https://en.wikipedia.org/wiki/Symbiogenesis
5 BBC Radio 4, The Infinite Monkey Cage, the wood wide web www.bbc.co.uk/sounds/play/m00193mg
6 https://donellameadows.org/archives/dancing-with-systems
7 www.abundantearth.coop
8 https://en.wikipedia.org/wiki/Viveka
9 Kahlil Gibran: *The Prophet*, Knopf Publishing, 1973

Least Change, pages 81-88

1 Bill Mollison: *Permaculture, A Designers' Manual*, Tagari Publications, 1988, p.15
2 https://knowledgebase.permaculture.org.uk/principles/make-least-change-greatest-possible-effect
3 www.permaculturenews.org/2020/03/26/on-making-the-least-change-for-the-greatest-effect https://arealgreenlife.com
4 https://evolution-music.co.uk
5 https://earthpercent.org
6 www.musicdeclares.net
7 Graham Bell: *The Permaculture Way*, Permanent Publications, 1992
8 https://en.wikipedia.org/wiki/Pareto_principle
9 https://en.m.wikipedia.org/wiki/Principle_of_least_effort
10 George Kingsley Zipf: *Human Behaviour and the Principle of Least Effort: An iItroduction to Human Ecology*, Addison-Wesley Press Inc., 1949
11 https://en.m.wikipedia.org/wiki/Principle_of_least_effort
12 https://en.m.wikipedia.org/wiki/Path_of_least_resistance
13 https://grammarist.com/phrase/path-of-least-resistance-and-line-of-least-resistance
14 https://en.wikipedia.org/wiki/Pierre_de_Fermat
15 https://en.m.wikipedia.org/wiki/Desire_path
16 www.apricotcentre.co.uk
17 BBC Radio 4: Just One Thing with Michael Mosley, Think Yourself Stronger www.bbc.co.uk/sounds/play/m0010pn6
18 https://en.wikipedia.org/wiki/Twelve_leverage_points

19 Donella H. Meadows et al: *The Limits to Growth*, Universe Books, 1972
20 www.youtube.com/watch?v=MS-XKMN95dk
21 www.academyforchange.org/2019/12/07/leverage-points-iceberg-model-economic-development
22 https://donellameadows.org/archives/leverage-points-places-to-intervene-in-a-system

Limiting Factors, pages 89-97

1 www.permaculture.org.uk/index.php/design-methods/limiting-factors-removal
2 Aranya: *Permaculture Design: A step-by-step guide*, Permanent Publications, 2012
3 Chris Evans and Jakob Jespersen: *The Farmers' Handbook*, Format Printing Press, 2001
4 Ian L. McHarg: *Design with Nature*, The Natural History Press, 1969
5 www.self-willed-land.org.uk/permaculture/design_methods.htm
6 https://en.m.wikipedia.org/wiki/Limiting_factor
7 https://en.wikipedia.org/wiki/Carrying_capacity
8 https://en.m.wikipedia.org/wiki/Liebig%27s_law_of_the_minimum
9 www.rhs.org.uk/lawns/creating-wildflower-meadows
10 www.waldeneffect.org/blog/Limiting_factors_in_the_garden
11 https://drive.google.com/file/d/1zIw7UgYOvdPcNFbuhPtRqGGzMtEWOd8A/view?usp=sharing
12 https://en.m.wikipedia.org/wiki/Bottleneck
13 Ian L. McHarg: *Design with Nature*, The Natural History Press, 1969
14 https://medium.com/age-of-awareness/visionaries-of-regenerative-design-iv-ian-l-mcharg-1920-2001-ea6da90b1958
15 www.permaculture.org.uk/design-methods/mcharg-exclusion-method
16 www.ou.edu/class/webstudy/fehler/E3/go/introduction.html
17 Looby Macnamara: *People & Permaculture*, Permanent Publications, 2012
18 www.thinklikeatree.co.uk
19 Sarah Spencer: *Think like a Tree: The Natural Principles Guide to Life*, Swarkestone Press, 2019
20 https://en.wikipedia.org/wiki/Parkinson%27s_law
21 www.smartertransport.uk/does-building-more-roads-reduce-congestion
22 https://muurgedichten.nl/en/muurgedicht/natur-und-kunst-1802
23 https://en.wikipedia.org/wiki/Oblique_Strategies
24 https://linktr.ee/queensscreech

Observe and Interact, pages 98-106

1 Bill Mollison: *Permaculture: A Designers' Manual*, Tagari Publications, 1988
2 David Holmgren: *Permaculture: Principles and Pathways Beyond Sustainability*, Melliodora, 2002
3 David Holmgren: *Permaculture: Principles and Pathways Beyond Sustainability*, Melliodora, 2002, p.15
4 Bill Mollison: *Permaculture: A Designers' Manual*, Tagari Publications, 1988, p.ix
5 www.weforum.org/agenda/2017/01/humans-have-more-than-5-senses/#:~:text=Neuroscientists
6 Lusi Alderslowe, Gaye Amus and Didi A. Devapriya: *Earth Care, People Care and Fair Share in Education, The Children in Permaculture Manual*, Eco-Logic Books, 2018
7 https://en.wikipedia.org/wiki/Lemon_balm
8 Patrick Whitefield: *The Living Landscape: How to Read and Understand It*, Permanent Publications, 2009
9 www.youtube.com/watch?v=MS-XKMN95dk
10 BBC Radio 4: The Senses, Synaesthesia: when senses merge www.bbc.co.uk/sounds/play/m000169z
11 https://en.wikipedia.org/wiki/Confirmation_bias
12 www.mrc-cbu.cam.ac.uk/people/matt.davis/cmabridge
13 Squares puzzle: the official answer is 30 (1 big square of 4 × 4 units, 4 squares of 3 × 3 units, 9 squares of 2 × 2 units, and 16 1 × 1 squares). This can also be demonstrated through the sum $1^2 + 2^2 + 3^2 + 4^2$
14 www.youtube.com/watch?v=Ikb5FCZPhEU
15 https://soilsoulstory.medium.com/decolonizing-permaculture-with-principle-0-9c027e4726c1
16 https://permacultureprinciples.com/permaculture-principles/_1
17 www.permaculture.co.uk/issue/winter-2021
18 www.youtube.com/watch?v=Yos6TKTbfSs
19 https://lizpostlethwaite.co.uk/2022/07/02/observe-and-interact
20 https://sawa-architecture.org/projects/a-house-for-a-victim

Patterns and Details, pages 107-114

1 Bill Mollison: *Permaculture: A Designers' Manual*, Tagari Publications, 1988
2 Bill Mollison with Reny Mia Slay: *Introduction to Permaculture*, Tagari, 1991
3 David Holmgren: *Permaculture: Principles and Pathways Beyond Sustainability*, Melliodora, 2002
4 www.freepermaculture.com/principles
5 https://kindredmedia.org/2011/12/social-permaculture-by-starhawk
6 Sarah Spencer: *Think like a Tree: The Natural Principles Guide to Life*, Swarkestone Press, 2019
7 Looby Macnamara: *People & Permaculture*, Permanent Publications, 2012
8 David Holmgren: *Permaculture: Principles and Pathways Beyond Sustainability*, Melliodora, 2002, p.xix
9 https://permacultureprinciples.com/permaculture-principles/_7
10 www.leadershipcentre.org.uk/wp-content/uploads/2016/02/The-Art-of-Change-Making.pdf p.23
11 BBC Radio 4: A Year to Change Your Mind by Dr Lucy Maddox, Episode 5: May www.bbc.co.uk/sounds/play/p0fslwwf
12 Priya Hemenway: *The Secret Code: The Mysterious Formula that Rules Art, Nature and Science*, Evergreen, 2008
13 https://en.wikipedia.org/wiki/Apophenia
14 https://lizpostlethwaite.co.uk/2022/10/22/

learning-from-patterns-in-nature
15 Bill Mollison: *Permaculture: A Designers' Manual*, Tagari Publications, 1988, p.71
16 Donella H. Meadows: *Thinking in Systems: A Primer*, Chelsea Green Publishing, 2008, p.11
17 https://en.wikiquote.org/wiki/Kurt_Lewin
18 Christopher Alexander et al: *A Pattern Language: Towns, Buildings, Construction*; Oxford University Press, 1977
19 Christopher Alexander et al: *A Pattern Language: Towns, Buildings, Construction*; Oxford University Press, 1977, p.xiii
20 www.permaculturewomen.com/scale-of-permanence
21 Dave Jacke and Eric Toensmeier: *Edible Forest Gardens: 2 Volume Set*, Chelsea Green Publishing Co., 2005
22 Adam Brock: *Change Here Now: Permaculture Solutions for Personal and Community Transformation*, North Atlantic Books, 2017
23 www.transitionculture.org/2010/06/04/rethinking-transition-as-a-pattern-language-an-introduction/
24 https://web.archive.org/web/20220616050109 www.patterns.transitionresearchnetwork.org
25 https://groupworksdeck.org
26 https://groupworksdeck.org/patterns/Breaking_Bread_Together
27 Bill Mollison with Reny Mia Slay: *Introduction to Permaculture*, Tagari, 1991, p.67
28 BBC Radio 4: Uncharted with Hannah Fry, The Returning Soldier www.bbc.co.uk/programmes/m001qw4s
29 https://en.wikipedia.org/wiki/Gaia_hypothesis
30 BBC Radio 4: Uncharted with Hannah Fry, The Hockey Stick www.bbc.co.uk/programmes/m001qw93

The Problem is the Solution, pages 115-120

1 Bill Mollison: *Permaculture: A Designers' Manual*, Tagari Publications, 1988
2 Bill Mollison with Reny Mia Slay: *Introduction to Permaculture*, Tagari, 1991, p.30
3 Bill Mollison: *Permaculture: A Designers' Manual*, Tagari Publications, 1988, p.15
4 Rosemary Morrow: *Earth User's Guide to Permaculture*, Kangaroo Press, 1993
5 Graham Burnett: *Permaculture: A Beginner's Guide*, Permanent Publications, 2001
6 www.youtube.com/watch?v=MS-XKMN95dk
7 www.permaculture.org.uk/sites/default/files/page/document/core_curriculum_3.0_peer_reviewed2020.pdf
8 David Holmgren: *Permaculture: Principles and Pathways Beyond Sustainability*, Melliodora, 2002, p.15
9 https://en.wikipedia.org/wiki/Herbivore_adaptations_to_plant_defense
10 https://en.wikipedia.org/wiki/Plant_defense_against_herbivory
11 www.pbs.org/lifeofbirds/brain
12 www.youtube.com/watch?v=g83FuoR2GGM
13 www.rokhak.com
14 https://saigoneer.com/saigon-environment/24292-an-ode-to-water-hyacinth,-vietnam-s-invasive,-beautiful-aquatic-plant
15 www.youtube.com/watch?v=k8fsVzyj-PA
16 https://online.hbs.edu/blog/post/growth-mindset-vs-fixed-mindset
17 www.one-eighty.org/news/the-importance-of-surrounding-yourself-with-positivity
18 https://youtu.be/u5um8QWWRvo?si=PNcmi2Fu_g9fvxTa
19 https://en.wikipedia.org/wiki/Critical_thinking
20 https://kanbanize.com/lean-management/improvement/5-whys-analysis-tool
21 www.debono.com/de-bono-thinking-lessons-1/1.-PMI-lesson-plan
22 https://en.wikipedia.org/wiki/SWOT_analysis
23 https://grasscrete.com/grasscrete

Produce No Waste, pages 121-127

1 Bill Mollison: *Permaculture: A Designers' Manual*, Tagari Publications, 1988, p.18
2 Graham Burnett: *Permaculture: A Beginner's Guide*, Permanent Publications, 2001
3 Bill Mollison with Reny Mia Slay: *Introduction to Permaculture*, Tagari, 1991
4 https://permacultureprinciples.com/permaculture-principles/_6
5 www.freepermaculture.com/principles
6 Sarah Spencer: *Think like a Tree: The Natural Principles Guide to Life*, Swarkestone Press, 2019, p.109
7 www.patchoftheplanet.co.uk/diplomaportfolio
8 www.dcw.co.uk/the-history-of-rubbish-removal-and-recycling-services
9 BBC Radio 4: The Compass, The History of Wastefulness www.bbc.co.uk/sounds/play/w3csy52r
10 https://lizpostlethwaite.co.uk/2022/02/09/produce-no-waste-a-creative-perspective/
11 www.youtube.com/watch?v=V7C6woDdEtA
12 https://viablealternativenergy.com/waste-hierarchy
13 www.fairphone.com/en
14 www.actiononplastic.org/resources
15 BBC Radio 4, Rubbish: The Great Waste Crisis www.bbc.co.uk/sounds/play/b03zy4hn
16 www.youtube.com/watch?v=hmGrI_BVlnc

Relative Location, pages 128-135

1 John Quinney article *Mother Earth News* 1st July 1984 www.motherearthnews.com/homesteading-and-livestock/permaculture-design-part-1-zmazz84jazloeck
2 Bill Mollison with Reny Mia Slay: *Introduction to Permaculture*, Tagari, 1991
3 Sego Jackson *Living with the Land* article winter 1984 www.context.org/iclib/ic08/jackson
4 www.permaculture.org.uk/principles/efficient-energy-planning-zone-sector-and-slope
5 https://en.wikipedia.org/wiki/Johann_Heinrich_von_Th%C3%BCnen
6 Charles Sale: *The Specialist*, Putnam & Company Ltd, 1930

7 Patrick Whitefield: *Permaculture in a Nutshell*, Permanent Publications, 1993
8 Chris Dixon, transcript of David Holmgren course www.konsk.co.uk/resource/holm2.htm
9 Bill Mollison: *Permaculture: A Designers' Manual*, Tagari Publications, 1988, p.47

Renewable Resources and Services, pages 136-142

1 Bill Mollison: *Permaculture: A Designers' Manual*, Tagari Publications, 1988
2 Bill Mollison: *Permaculture: A Designers' Manual*, Tagari Publications, 1988, p.17
3 Bill Mollison with Reny Mia Slay: *Introduction to Permaculture*, Tagari, 1991
4 www.motherearthnews.com/homesteading-and-livestock/permaculture-design-part-1-zmazz84jazloeck/
5 www.yorkshirewillow.co.uk/willow-spiling-riverbank-stabilisation
6 David Holmgren: *Permaculture: Principles and Pathways Beyond Sustainability*, Melliodora, 2002
7 David Holmgren: *Permaculture: Principles and Pathways Beyond Sustainability*, Melliodora, 2002, p.94
8 https://permacultureprinciples.com/permaculture-principles/_5
9 https://projects.sare.org/wp-content/uploads/1346EW94-009.001.pdf
10 https://savoursoilpermaculture.com/wp-content/uploads/2021/10/the-principles-of-permaculture-bill-mollison-david-holmgren-.pdf
11 E. F. Schumacher: *Small is Beautiful: A Study of Economics as if People Mattered*, Blond and Briggs, 1973
12 https://en.wikipedia.org/wiki/Appropriate_technology
13 https://en.wikipedia.org/wiki/Universal_nut_sheller
14 Sarah Spencer: *Think like a Tree: The Natural Principles Guide to Life*, Swarkestone Press, 2019, p.113
15 http://isleofeigg.org/eigg-electric
16 www.sarahl.com
17 https://transitionnetwork.org
18 www.wonderbagworld.com
19 https://media2-production.mightynetworks.com/asset/12582477/Permaculture_principles-Starhawk.pdf
20 David Holmgren: *Permaculture: Principles and Pathways Beyond Sustainability*, Melliodora, 2002, p.99

Resilience, pages 143-149

1 https://docs.google.com/document/d/1BOvt8HrXsPVtOSLLKhTH4hEtY7eyfebr/
2 https://media2-production.mightynetworks.com/asset/12582477/Permaculture_principles-Starhawk.pdf
3 Sarah Spencer: *Think like a Tree: The Natural Principles Guide to Life*, Swarkestone Press, 2019, p149-180
4 www.rootsnpermaculture.com
5 https://en.wikipedia.org/wiki/Ecological_resilience
6 https://youtu.be/NWH8N-BvhAw
7 Major General Paul Nansen: *Stand up Straight: 10 Life Lessons from the Royal Military Academy Sandhurst*, Century, 2019
8 www.abundantearth.coop
9 www.resilience.org/stories/2016-02-24/growing-the-growers-communication-and-wellbeing-on-farms

Self Regulate, pages 150-157

1 David Holmgren: *Permaculture: Principles and Pathways Beyond Sustainability*, Melliodora, 2002
2 https://en.wikipedia.org/wiki/Stoicism
3 Bill Mollison: *Permaculture: A Designers' Manual*, Tagari Publications, 1988
4 https://en.wikipedia.org/wiki/Gaia_hypothesis
5 www.lse.ac.uk/granthaminstitute/explainers/what-is-climate-change-legislation
6 Bill Mollison: *Permaculture: A Designers' Manual*, Tagari Publications, 1988, p.1
7 https://en.wikipedia.org/wiki/Positive_feedback
8 https://en.wikipedia.org/wiki/Negative_feedback
9 https://en.wikipedia.org/wiki/Positive_feedback#/media/File:20220726_Feedbacks_affecting_global_warming_and_climate_change_-_block_diagram.svg
10 https://en.wikipedia.org/wiki/Kaibab_Plateau
11 https://en.wikipedia.org/wiki/Transtheoretical_model
12 https://medium.com/centre-for-public-impact/what-gets-measured-gets-managed-its-wrong-and-drucker-never-said-it-fe95886d3df6
13 www.small-improvements.com/blog/nonviolent-feedback
14 https://sergiocaredda.eu/organisation/tools/models-the-lippitt-knoster-model-for-managing-complex-change
15 www.youtube.com/watch?v=OfT8zKsMqzo

Small and Slow, pages 158-165

1 https://executivecoachinglondon.com/life-choices/pace-of-modern-life
2 https://permacultureprinciples.com/permaculture-principles/_9
3 Bill Mollison with Reny Mia Slay: *Introduction to Permaculture*, Tagari, 1991
4 www.context.org/iclib/ic28/mollison
5 Rosemary Morrow: *Earth User's Guide to Permaculture*, 2nd edition: Permanent Publications, 2006
6 www.youtube.com/watch?v=MS-XKMN95dk
7 https://en.wikipedia.org/wiki/The_Tortoise_and_the_Hare
8 Helen Ward: *The Hare and the Tortoise*, Templar Publishing, 2000
9 www.idler.co.uk
10 https://en.wikipedia.org/wiki/Slow_Food
11 https://en.wikipedia.org/wiki/Cittaslow
12 www.tate.org.uk/tate-etc/issue-46-summer-2019/slow-art-take-time-jonathan-p-watts
13 www.bbc.co.uk/sounds/brand/p05k5bq0

14 https://en.wikipedia.org/wiki/Slow_television
15 https://drive.google.com/file/d/1dR3WgfrRsEe6rDFYQWQyPAoTE5JSu20X/view?usp=sharing
16 www.gabebc.com/a-point-just-passed
17 www.amphora.ca/ideas/2020/6/26/cheap-fast-or-good-you-can-pick-two-but-not-all-three
18 E. F. Schumacher: *Small is Beautiful: A Study of Economics as if People Mattered*, Blond and Briggs, 1973
19 https://en.m.wikipedia.org/wiki/Leopold_Kohr
20 https://centerforneweconomics.org/newsletters/the-heart-of-the-problem
21 https://en.wikipedia.org/wiki/Flow_(psychology)
22 www.nytimes.com/2004/12/28/science/bigger-is-better-view-of-evolution-gains-credence.html
23 www.bayoakomolafe.net/post/the-times-are-urgent-lets-slow-down
24 https://pacific-edge.info/2015/02/11/small-and-slow-solutions-a-principle-too-little-too-late/
25 www.youtube.com/watch?v=pAoZWyLMb6c
26 www.thinklikeatree.co.uk

Stacking and Succession, pages 166-172

1 Bill Mollison: *Permaculture Two: Practical Design for Town and Country in Permanent Agriculture*, Tagari Publications, 1979
2 https://en.wikipedia.org/wiki/Ecological_succession
3 https://tobyhemenway.com/resources/ethics-and-principles
4 www.nature.com/articles/d42859-019-00010-6
5 Bill Mollison with Reny Mia Slay: *Introduction to Permaculture*, Tagari, 1991
6 https://en.m.wikipedia.org/wiki/Robert_Hart_(horticulturist)
7 https://en.m.wikipedia.org/wiki/Forest_gardening
8 https://en.m.wikipedia.org/wiki/Toyohiko_Kagawa
9 J. Russell Smith: *Tree Crops: A Perennial Agriculture*, Harcourt Brace and Company, 1929
10 https://en.m.wikipedia.org/wiki/J._Russell_Smith
11 https://en.m.wikipedia.org/wiki/Agroforestry

Work with Nature, pages 173-181

1 www.spiritoftheland.co.uk/section855491_762896.html
2 Bill Mollison: *Permaculture Two: Practical Design for Town and Country in Permanent Agriculture*, Tagari Publications, 1979
3 https://en.m.wikipedia.org/wiki/Masanobu_Fukuoka
4 Masanobu Fukuoka: *One Straw Revolution*, Rodale Press, 1975
5 Bill Mollison: *Permaculture Two: Practical Design for Town and Country in Permanent Agriculture*, Tagari Publications, 1979, p.1
6 Bill Mollison: *Permaculture: A Designers' Manual*, Tagari Publications, 1988
7 www.context.org/iclib/ic42/jeavons
8 Patrick Whitefield: *The Living Landscape: How To Read and Understand It*, Permanent Publications, 2009
9 https://media2-production.mightynetworks.com/asset/12582477/Permaculture_principles-Starhawk.pdf
10 Patrick Whitefield: *Permaculture in a Nutshell*, Permanent Publications, 1993
11 Patrick Whitefield: *Permaculture in a Nutshell*, Permanent Publications, 1993, p.14
12 www.bbc.co.uk/sounds/play/b007w2w3
13 https://nvc-uk.com/2018/03/05/certification-candidate-stumbles-upon-self-acceptance

Yield Unlimited, pages 182-187

1 David Holmgren: *Permaculture: Principles and Pathways Beyond Sustainability*, Melliodora,, 2002
2 Bill Mollison: *Permaculture: A Designers' Manual*, Tagari Publications, 1988
3 Rosemary Morrow: *Earth User's Guide to Permaculture*, Kangaroo Press, 1993
4 Heather C. Flores: *Food Not Lawns*, Chelsea Green Publishing Co., 2006
5 Toby Hemenway: *Gaia's Garden*, 2nd edition, Chelsea Green Publishing Co., 2009
6 www.youtube.com/watch?v=MS-XKMN95dk
7 www.abc.net.au/everyday/crme-brule-flavoured-with-jerusalem-artichoke-and-white-truffle-/8935062
8 https://miro.com/app/board/o9J_ldjcSak=

How I Used the Principles to Write this Book, pages 188-194

1 https://greendurham.org.uk
2 Looby Macnamara: *Cultural Emergence*, Permanent Publications, 2020

Relevant Books

Books that inspired and developed the permaculture principles in chronological order

J. Russell Smith: *Tree Crops: A Perennial Agriculture,* Harcourt Brace and Company, 1929

Ian L. McHarg: *Design with Nature,* The Natural History Press, 1969

Donella H. Meadows et al: *The Limits to Growth,* Universe Books, 1972

Kenneth E. F. Watt: *Principles of Environmental Science,* McGraw-Hill, 1973

E. F. Schumacher: *Small is Beautiful: A Study of Economics as if People Mattered,* Blond and Briggs, 1973

Masanobu Fukuoka: *One Straw Revolution,* Rodale Press, 1975

Christopher Alexander et al: *A Pattern Language: Towns, Buildings, Construction,* Oxford University Press, 1977

Bill Mollison and David Holmgren: *Permaculture One: A Perennial Agriculture for Human Settlements,* Tagari Publications, 1978

Bill Mollison: *Permaculture Two: Practical Design for Town and Country in Permanent Agriculture,* Tagari Publications, 1979

James Lovelock: *Gaia: A New Look at Life on Earth,* Oxford University Press, 1979

Bill Mollison: *Permaculture Design Course Series,* Yankee Permaculture, 1981

Bill Mollison: *Permaculture: A Designers' Manual,* Tagari Publications, 1988

Bill Mollison with Reny Mia Slay: *Introduction to Permaculture,* Tagari Publications, 1991

Graham Bell: *The Permaculture Way,* Permanent Publications, 1992

Rosemary Morrow: *Earth User's Guide to Permaculture,* Kangaroo Press, 1993

Patrick Whitefield: *Permaculture in a Nutshell,* Permanent Publications, 1993

Ross Mars: *The Basics of Permaculture Design,* Candlelight Trust, 1996

Robert A. de J. Hart: *Forest Gardening: Rediscovering Nature and Community in a Post-Industrial Age,* Green Books, 1996

Sim Van Der Rym and Stuart Cowen: *Ecological Design,* Island Press, 1996

Graham Burnett: *Permaculture: A Beginner's Guide,* Permanent Publications, 2001

Chris Evans and Jakob Jespersen: *The Farmers' Handbook,* Format Printing Press, 2001

David Holmgren: *The Essence of Permaculture Principles,* Melliodora, 2001

David Holmgren: *Permaculture: Principles and Pathways Beyond Sustainability,* Melliodora, 2002

Patrick Whitefield: *The Earth Care Manual: A Permaculture Handbook for Britain and Other Temperate Climates,* Permanent Publications, 2004

Dave Jacke and Eric Toensmeier: *Edible Forest Gardens: 2 Volume Set*, Chelsea Green Publishing Co., 2005
Rosemary Morrow: *Earth User's Guide to Permaculture*, 2nd edition, Permanent Publications, 2006
Heather Flores: *Food Not Lawns*, Chelsea Green Publishing Co., 2006
Ian Lillington: *The Holistic Life: Sustainability through Permaculture*, Axiom, 2007
Donella H. Meadows: *Thinking in Systems: A Primer*, Chelsea Green Publishing Co., 2008
Patrick Whitefield: *The Living Landscape: How To Read and Understand It*, Permanent Publications, 2009
Toby Hemenway: *Gaia's Garden*, 2nd edition, Chelsea Green Publishing Co., 2009
Sepp Holzer: *Sepp Holzer's Permaculture: A Practical Guide to Small-Scale, Integrative Farming and Gardening*, Permanent Publications, 2010
Looby Macnamara: *People & Permaculture*, Permanent Publications, 2012
Aranya: *Permaculture Design: A Step-by-Step Guide*, Permanent Publications, 2012
Tomas Remiarz: *Forest Gardening in Practice; An Illustrated Practical Guide for Homes, Communities and Enterprises*, Permanent Publications, 2017
Adam Brock: *Change Here Now: Permaculture Solutions for Personal and Community Transformation*, North Atlantic Books, 2017
Lusi Alderslowe, Gaye Amus and Didi A. Devapriya: *Earth Care, People Care and Fair Share in Education, The Children in Permaculture Manual*, Eco-Logic Books, 2018
Sarah Spencer: *Think like a Tree: The Natural Principles Guide to Life*, Swarkestone Press, 2019
Looby Macnamara: *Cultural Emergence*, Permanent Publications, 2020
Delvin Solkinson and Grace Solkinson: *Permaculture Design Notes*, Dew Drop Press, 2021
Rosemary Morrow: *Earth Restorer's Guide to Permaculture*, Melliodora, 2022
Kirsten Bradley: *The Milkwood Permaculture Living Handbook*, Murdoch Books, 2023

Massive Thanks

To all of the collaborators and contributors to the various chapters including Andy Goldring, Aranya, Arnaud Fache, Barry Jones, Beth Silverbirch, Carla Moss, Cathrine Dolleris, Chris Evans, Chris Warburton-Brown, Delvin Solkinson, Emma Leaf-Grimshaw, Ed Tyler, Graham Bell, Hannah Thorogood, Helen White, Ian Lillington, James Taylor, Jenni Brooks, Jen Rouse, Jo Barker, Jo Holleran, Katie Shepherd, Keira Oliver, Liz Postlethwaite, Looby Macnamara, Lusi Alderslowe, Maddy Harland, Marina O'Connell, Mark Shipperlee, Neil Kingsnorth, Peter Cow, Pia Castleton, Rachel Phillips, Rakesh Rootsman Rak, Ruth Starr-Keddle, Sally Hughes, Sam Woods, Sarah Spencer, Stephen Andrews, Steve Charter, Tobias Sonrisa Sturmer, Tom Henfrey, Tomas Remiarz and Wenderlynn Bagnall.

To everyone that has contributed to the development of the permaculture principles over many decades through books, articles, handouts, drawings and websites, all of which have been immensely helpful in my research, including Andy Goldring, Angelo Eliades, Aranya, Bill Mollison, Brett Pritchard, Chris Dixon, Chris Evans, Colleen Stevenson, Dan Palmer, David Holmgren, Delvin Solkinson, Doug Crouch, Graham Bell, Graham Burnett, Heather Jo Flores, Ian Lillington, John Quinney, Kate Martignier, Kirsten Bradley, Larry Santoyo, Lisa DePiano, Liz Postlethwaite, Looby Macnamara, Maddy Harland, Masanobu Fukuoka, Meg McGowan, Mirranda Burton, Morag Gamble, Neil Kingsnorth, Patricia Allison, Patrick Whitefield, Rakesh Rootsman Rak, Reny Mia Slay, Rob Hopkins, Robin Clayfield, Rosemary Morrow, Ross Mars, Russ Grayson, Sally Hughes, Sandy Cruz, Sarah Queblatin, Sarah Spencer, Sepp Holzer, Starhawk, Toby Hemenway and Tomas Remiarz.

Index

permaculture
Transforming Culture
starting at home
permaculture
Everyday Activism
permaculture
WHILE THE WORLD WAITS,
WE REGENERATE